Energy for Water

Water and energy are inextricably linked as unsound management of either resource can have an impact on the cost, availability, and sustainability of the other. This book explores the "energy for water" component of the water–energy nexus. It offers diverse case studies from around the world including the deserts of Saudi Arabia, rural China, Pakistan's Indus Basin, arid Greek islands, and urban centers such as Los Angeles. The analyses show that while many regions face unique water scarcity challenges, they are all united by the fact that solutions require mobilizing energy.

This book focuses on how different policies and technologies are changing the way societies use energy to extract, treat, and transport water. In terms of policy, chapters explore how initiatives aimed at reducing demand for water and improved integrated resource planning can lead to energy savings. Regarding technology, case studies highlight the pros and cons of different methods of meeting water demand.

Through exploring both policy and technology across a wide range of diverse case studies, the book offers a robust explanation of the "energy for water" side of the water–energy nexus equation, making it valuable reading for academics and policymakers.

This book was originally published as a special issue in the *International Journal of Water Resources Development.*

Christopher Napoli is a research fellow at the King Abdullah Petroleum Studies and Research Center, Riyadh, Saudi Arabia, where he researches and advises on natural resource economics and environmental management. Prior to this, he was Assistant Professor of International Political Economy at the University of Nottingham, Malaysia Campus. He has a PhD from the University of Kent, UK, and a Bachelor of Commerce degree from Queen's University, Canada.

Routledge Special Issues on Water Policy and Governance

www.routledge.com/series/WATER

Edited by
Cecilia Tortajada (IJWRD), Third World Centre for Water Management, Mexico
James Nickum (WI), International Water Resources Association, France

Most of the world's water problems, and their solutions, are directly related to policies and governance, both specific to water and in general. Two of the world's leading journals in this area, the *International Journal of Water Resources Development* and *Water International* (the official journal of the International Water Resources Association), contribute to this special issues series, aimed at disseminating new knowledge on the policy and governance of water resources to a very broad and diverse readership all over the world. The series should be of direct interest to all policy makers, professionals and lay readers concerned with obtaining the latest perspectives on addressing the world's many water issues.

Water for Food Security
Challenges for Pakistan
Edited by Claudia Ringler and Arif Anwar

Water Management and Climate Change
Dealing with uncertainties
Edited by Cecilia Tortajada, Asit K. Biswas and Avinash Tyagi

Integrated Water Resources Management
From concept to implementation
Edited by Cecilia Tortajada

Water Infrastructure
Edited by Cecilia Tortajada and Asit K. Biswas

Frontiers of Land and Water Governance in Urban Regions
Edited by Thomas Hartmann and Tejo Spit

The Water-Energy-Food Nexus in the Middle East and North Africa
Edited by Martin Keulertz and Eckart Woertz

Sustainability in the Water-Energy-Food Nexus
Edited by Anik Bhaduri, Claudia Ringler, Ines Dombrowsky, Rabi Mohtar and Waltina Scheumann

Transboundary Water Cooperation
Principles, practice and prospects for China and its neighbours
Edited by Patricia Wouters, Huiping Chen and James E. Nickum

Water Reuse Policies for Potable Use
Edited by Cecilia Tortajada and Choon Nam Ong

Hydrosocial Territories and Water Equity
Theory, governance and sites of struggles
Edited by Rutgerd Boelens, Ben Crow, Jaime Hoogesteger, Flora E. Lu, Erik Swyngedouw and Jeroen Vos

Energy for Water
Regional case studies
Edited by Christopher Napoli

Energy for Water

Regional Case Studies

Edited by
Christopher Napoli

Routledge
Taylor & Francis Group
LONDON AND NEW YORK

KAPSARC
مركز الملك عبدالله للدراسات والبحوث البترولية
King Abdullah Petroleum Studies and Research Center

First published 2018 by Routledge

2 Park Square, Milton Park, Abingdon, Oxfordshire OX14 4RN
52 Vanderbilt Avenue, New York, NY 10017

Routledge is an imprint of the Taylor & Francis Group, an informa business

First issued in paperback 2019

British Library Cataloguing in Publication Data
A catalogue record for this book is available from the British Library

ISBN 13: 978-1-138-74901-6 (hbk)
ISBN 13: 978-0-367-23408-9 (pbk)

Typeset in Myriad Pro
by diacriTech, Chennai

Publisher's Note
The publisher accepts responsibility for any inconsistencies that may have arisen during the conversion of this book from journal articles to book chapters, namely the possible inclusion of journal terminology.

Contents

Citation Information

The chapters in this book were originally published in the *International Journal of Water Resources Development*, volume 32, issue 3 (May 2016). When citing this material, please use the original page numbering for each article, as follows:

Chapter 1
A framework for understanding energy for water
Christopher Napoli and Berenice Garcia-Tellez
International Journal of Water Resources Development, volume 32, issue 3 (May 2016) pp. 339–361

Chapter 2
The energy trade-offs of adapting to a water-scarce future: case study of Los Angeles
Kelly Twomey Sanders
International Journal of Water Resources Development, volume 32, issue 3 (May 2016) pp. 362–378

Chapter 3
Quantifying and managing urban water-related energy use systemically: case study lessons from Australia
Steven John Kenway and Ka Leung Lam
International Journal of Water Resources Development, volume 32, issue 3 (May 2016) pp. 379–397

Chapter 4
Energy for freshwater supply, use and disposal in the Netherlands: a case study of Dutch households
P.W. Gerbens-Leenes
International Journal of Water Resources Development, volume 32, issue 3 (May 2016) pp. 398–411

Chapter 5

Evaluating the economic viability of solar-powered desalination: Saudi Arabia as a case study
Christopher Napoli and Bertrand Rioux
International Journal of Water Resources Development, volume 32, issue 3 (May 2016) pp. 412–427

Chapter 6

Responding to water challenges in Greece through desalination: energy considerations
A. Kartalidis, E. Tzen, E. Kampragkou and D. Assimacopoulos
International Journal of Water Resources Development, volume 32, issue 3 (May 2016) pp. 428–441

Chapter 7

Modernization of irrigation systems in Spain: review and analysis for decision making
Javier Alarcón, Alberto Garrido and Luis Juana
International Journal of Water Resources Development, volume 32, issue 3 (May 2016) pp. 442–458

Chapter 8

Drivers of groundwater use and technical efficiency of groundwater, canal water, and conjunctive use in Pakistan's Indus Basin Irrigation System
Dawit Mekonnen, Afreen Siddiqi and Claudia Ringler
International Journal of Water Resources Development, volume 32, issue 3 (May 2016) pp. 459–476

Chapter 9

Energy for water utilization in China and policy implications for integrated planning
Xi Li, Jie Liu, Chunmiao Zheng, Guoyi Han and Holger Hoff
International Journal of Water Resources Development, volume 32, issue 3 (May 2016) pp. 477–494

Notes on Contributors

Javier Alarcón is based at the ETS Ingenieros Agrónomos Research Centre for the Management of Agricultural and Environmental Risks (CEIGRAM), Technical University of Madrid, Spain.

Dionysis Assimacopoulos is a Professor at the School of Chemical Engineering, National Technical University of Athens, Greece. His research projects include work in water resource management, energy management, and environmental management.

Berenice Garcia-Tellez is a research assistant at the King Abdullah Petroleum Studies and Research Center, Riyadh, Saudi Arabia.

Alberto Garrido is Professor of Agricultural and Natural Resource Economics at the ETS Ingenieros Agrónomos Research Centre for the Management of Agricultural and Environmental Risks (CEIGRAM), Technical University of Madrid, Spain. His research areas include natural resource economics and management, risk management, and agricultural policy.

P.W. Gerbens-Leenes is a researcher at the Water Management Group, Twente Water Centre, University of Twente, Enschede, the Netherlands. Her research is concerned with the relationship between water footprints and energy.

Guoyi Han is a research fellow at the Stockholm Environment Institute, Sweden.

Holger Hoff is a senior research fellow at the Stockholm Environment Institute, Sweden. His research interests include climate adaptation and integrated natural resource management.

Luis Juana is based at the ETS Ingenieros Agrónomos Research Centre for the Management of Agricultural and Environmental Risks (CEIGRAM), Technical University of Madrid, Spain.

Eleni Kampragkou is based at the School of Chemical Engineering, National Technical University of Athens, Greece.

Avraam Kartalidis is a graduate student at the School of Chemical Engineering, National Technical University of Athens, Greece.

Steven John Kenway is an associate professor at the School of Chemical Engineering, University of Queensland, Brisbane, Australia. His research interests include the urban water cycle, integrated water and energy planning, and urban water security.

Ka Leung Lam is a graduate student at the School of Chemical Engineering, University of Queensland, Brisbane, Australia. His principal area of research is into the energy implications of urban water options.

Xi Li is based at the Institute of Water Sciences and College of Engineering, Peking University, Beijing, China.

Jie Liu is an associate professor at the Institute of Water Sciences and College of Engineering, Peking University, Beijing, China. Her research areas include groundwater flow/transport modelling, integrated hydrologic modelling, and sustainable water management.

Dawit Mekonnen is a research fellow at the Environment and Production Technology Division, International Food Policy Research Institute, Washington, DC, USA. His research interests include productivity and efficiency analysis, water economics, agricultural input systems, and impact evaluation of development projects.

Christopher Napoli is a research fellow at the King Abdullah Petroleum Studies and Research Center, Riyadh, Saudi Arabia, where he researches and advises on natural resource economics and environmental management. Prior to this, he was Assistant Professor of International Political Economy at the University of Nottingham, Malaysia Campus.

Claudia Ringler is Deputy Division Director at the Environment and Production Technology Division, International Food Policy Research Institute, Washington, DC, USA. Her research interests are water resources management, including the use of river basin modelling, and natural resource policy focused at sustainable agricultural productivity growth.

Bertrand Rioux is a research associate at the King Abdullah Petroleum Studies and Research Center, Riyadh, Saudi Arabia.

Kelly Twomey Sanders is an assistant professor at the Department of Civil and Environmental Engineering, University of South California, Los Angeles, CA, USA. Her research seeks to increase the sustainable development of urban and agricultural systems.

Afreen Siddiqi is a research scientist at the Engineering Systems Division, Tata Center for Technology and Design, Massachusetts Institute of Technology, Cambridge, MA, USA. She is currently researching off-grid hydro energy systems in irregular channel flow for intensive agriculture in developing countries.

Eftihia Tzen is an independent engineer based in Athens, Greece.

Chunmiao Zheng is Professor of Hydrology and Water Resources at the Department of Geological Sciences, University of Alabama, Tuscaloosa, AL, USA, and at the School of Environmental Science and Engineering, South University of Science and Technology of China, Shenzhen, China. His research interests include studies of the impact of aquifer heterogeneities and preferential flow paths on contaminant transport, hydrogeology, and the management of water resources at watershed scales.

A framework for understanding energy for water

Christopher Napoli and Berenice Garcia-Tellez

King Abdullah Petroleum Studies and Research Center, Riyadh, Saudi Arabia

ABSTRACT
This article offers a framework for understanding how energy is used to meet water demand in countries. Specifically, the relationships between energy use and water scarcity, the location of renewable water resources, and aggregate water demand are explored. The article also examines how policy options such as water price reforms, agriculture subsidies and crop elimination may influence the energy use and energy intensity of water withdrawals. Conclusions suggest that while policy options exist, certain uncontrollable factors such as severe water scarcity or substantial freshwater abundance limit the ability of some countries to significantly improve the aggregate energy efficiency of water provision.

Introduction

Oceans cover 71% of the earth's surface and represent over 96% of the world's total water resources. Of the remaining water, half is frozen in glaciers or ice caps and a quarter is locked in saline underground aquifers. This leaves less than 1% of the world's total water resources available for consumption by life on earth, a small percentage in relative terms, but a figure representing 10.6 million km^3, or 1.5 million m^3 per person (own calculations, data from Gleick, 1993). As the hydrological cycle results in much of the world's freshwater being recycled after consumption, the current stock of water is more than enough to sustain life on earth.

While there is enough freshwater on earth, nearly 20% of the world's population faces physical water scarcity, a situation in which there is insufficient water to meet the demands of the ecosystem. The United Nations estimates that if population and economic growth continue to strain the world's water resources, roughly two-thirds of the world's population could be living in water-stressed regions by 2025 (United Nations Department of Economic & Social Affairs [UNDESA], 2014).

One of the fundamental components in addressing physical water scarcity is energy. This is because energy is a primary input for moving water from areas of relative abundance to areas of relative scarcity. For example, in California, water from the abundant north of the state travels hundreds of kilometres over the eastern Sierra Nevada mountain range in order to meet demand in Los Angeles, a region of increasing physical water scarcity. The energy required to transport water this distance is estimated to be 1.6 kWh/m^3, or three times the energy required to extract and treat water withdrawn from sources in Los Angeles (Wilkinson, 2000). The need for energy-intensive interbasin water transfers is expected to increase as the population and economy of the city grow.

In Jordan, a severely water-scarce country, a different sort of water transport occurs. Groundwater from aquifers located more than 1 km below the earth's surface is extracted for use by municipalities and agriculture. This extraction comes at a tremendous energy cost. Hayek (2014) estimates that the average energy required to extract groundwater in Jordan is roughly 1.2 kWh/m^3, or nine times the energy required to withdraw groundwater in the Netherlands, a country also reliant on groundwater, but with much shallower bores (Imperial Irrigation District, 2015). The case of Jordan is important because the country imports 97% of its energy from neighbouring countries, and these energy supplies are largely in the form of non-renewable fossil fuels that can experience large fluctuations in price (Talozi, Al Sakaji, & Altz-Stamm, 2015).

In rare cases, such as Spain, Australia and the Gulf region, seawater is desalinated, as neither interbasin transfers nor deep aquifer extraction are sufficient to meet water demand. Desalination represents the most energy-intensive option for meeting water demand, and often entails significant financial and environmental costs.

Exploring the relationship between energy and water demand is important for two reasons. First, a significant amount of energy is used for direct water services. For example, Sanders and Webber (2012) have estimated that 8.3% of 2010 annual primary energy consumption in the United States was used for the heating, chilling, treating, pressurizing and pumping of water by the municipal, industrial and agricultural sectors. Thus, water conservation can have a direct effect on energy use, which can help improve energy security and climate change objectives, particularly when fossil fuels are used as the energy source.

Second, while water and energy are inextricably linked, they are typically governed in silos, which can lead to policy fragmentation and a misalignment of incentives. In general, the energy sector considers water only as it relates to hydropower and water requirements for thermal power plants. Similarly, water planners are typically more concerned with supply augmentation and management, as opposed to energy issues (Malik, 2002). Improving coordination of the management of water and energy resources can increase efficiency in both areas. For example, Kumar, Scott, and Singh (2013) have shown that power tariff reform that includes pro rata pricing and higher unit rates for electricity would likely lead to improved efficiency and sustainability of groundwater use by farmers in India. This is an important consideration given that in 2010 India extracted 684 km^3 of water for agriculture, a figure representing 90% of total extractions (http://www.fao.org/nr/water/aquastat/data/query/).

This article offers a framework for understanding how energy is used to meet water demand in both water-scarce and water-abundant countries. Specifically, the supply and demand of water, and the energy required to withdraw that water, are disaggregated: supply is disaggregated by source and treatment process, while demand is disaggregated by user. This approach helps policy makers understand the ways decisions in one domain can affect the other, thus building the case for more integrated management of the water–energy nexus. Through a detailed examination of how energy is used for water supply, and how water is used in the economy, the framework also provides policy makers with a tool for assessing the possibilities (and limits) of efficiency improvements in both domains.

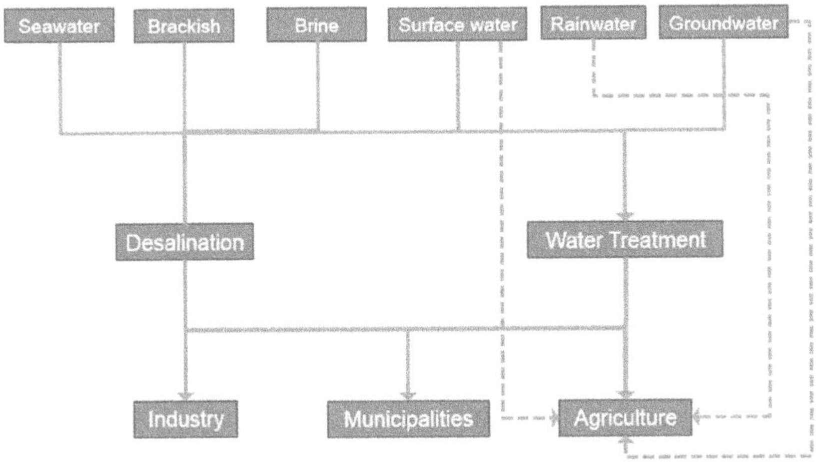

Figure 1. Sources and uses of water.

The article's conclusions are threefold. First, energy use for water is a function of not only water scarcity, but also the location of renewable water resources: countries with abundant renewable water resources that are located in deep aquifers, or far from demand centres, may still require substantial energy resources to meet water demand. Second, while economic solutions for managing water resources exist, they can be politically sensitive, and therefore difficult to implement. This often results in the overuse of water, which can lead to increases in both aggregate energy use and overall energy intensity of water extraction. Last, while some countries can reduce the energy used for water by eliminating water-intensive agricultural crops, certain uncontrollable factors, such as severe water scarcity or high water abundance, reduce the efficacy of this option. By disaggregating and exploring how supply and demand for water can affect energy use, this article offers a theoretical framework through which to examine the case studies on managing energy for water presented in this special issue.

The supply of water: implications for energy

As described in Figure 1, water can come from both conventional and unconventional sources. *Conventional sources* refers to surface water, groundwater and rainwater, all of which may be treated and then consumed, or sent directly from the source to the final consumer. *Unconventional sources* refers to seawater, brackish water, brine and wastewater, which must be purified by advanced treatment technologies prior to consumption.

The energy required to meet water demand is dependent on the type of water used, whether it is treated, and the technology used for treatment. In almost all cases, water used for agriculture is untreated, and thus consumes the least energy. For example, rainwater consumed directly by agriculture incurs effectively no energy footprint; and untreated surface water requires only minimal energy for extraction. As farmers begin to extract water from underground bores or use desalinated water, energy requirements increase.

Water extracted for municipal and industrial use is typically purified prior to consumption. This can increase energy requirements. Surface water is often contaminated and must be pumped through numerous filters and disinfected with chlorine and other chemicals before being distributed to the water grid. Groundwater, in contrast, is typically cleaner, and does not require much treatment aside from the addition of chlorine and other purifying chemicals. As a result, most of the energy required to withdraw and purify groundwater is used for extraction (Burton, 1996; Copeland, 2014; Plappally & Lienhard, 2012).

The energy required for unconventional water withdrawals is a function of three factors: the type of water withdrawn; the quantity of water withdrawn; and the desalination technology used. Two primary types of desalination technologies exist: thermal and membrane. Thermal desalination is a process in which saline water is vaporized, thus separating pure water from any salts, minerals and other contaminants (Tonner, 2008). Membrane desalination is a process whereby saline water is passed through one or more semipermeable membranes. The membranes separate pure water from salts and other impurities. The energy required for thermal desalination processes such as multistage flash distillation (MSF) and multiple-effect distillation (MED) is higher, and is independent of the salinity or the source of water. In contrast, the energy required for membrane technology, such as reverse osmosis (RO) and electrodialysis (ED), is generally less, and varies with the salinity of water: the more saline the water, the more energy-intensive the extraction and treatment. In addition to the specific technology used, other factors will affect the energy required for desalination, including output capacity of the plant, thermal design, membrane type (for membrane technologies), efficiency of the plant, and system configuration. The latter is important to consider for dual-purpose plants (i.e. plants designed for power and water production). While global online seawater desalination capacity increased significantly between 2000 and 2014, from 0.72 km^3 to 13.73 km^3 (Global Water Intelligence, 2015), given its high capital and energy costs, the technology contributes only a small fraction to overall water supplied.

Figure 2 describes how energy use increases with different water sources and extraction technologies. With the exception of rainfall, untreated surface water requires the least energy for extraction, with the variance in energy required due primarily to the varying efficiency of pumps for extraction. In emerging economies, where pumps are often less efficient and energy costs are lower, energy intensity is generally higher. The figure also shows that while some water sources are more energy-intensive than others, each source has a large range of possible energy intensities. For example, while it is generally assumed that treated wastewater is more energy-intensive than groundwater extraction, this may not be the case when groundwater bores are deep and pumps are inefficient. This is an important consideration in water-scarce countries that must rely on deep groundwater aquifers. Wastewater treatment may also offer a less energy-intensive alternative to desalination. In Saudi Arabia, a severely water-scarce country, the government has begun to encourage the reuse of wastewater, and plans have been put in place to expand wastewater collection and treatment systems to cover about 60% of urban areas by 2014 – up from 42% in 2010 (Ouda, 2014). To date, much of the recycled water has been used for urban-area landscaping and, to a lesser extent, crop irrigation (Ouda, 2014). It is hoped that wastewater will help alleviate stress on non-renewable aquifers through less energy-intensive, cheaper means than can be offered by desalination. The case of Saudi Arabia shows that when countries understand the economic and energy implications of different water sources, it is feasible for strategies to be implemented that minimize these costs.

An important source of water that is not considered in Figure 2 is water transported over long distances. Lack of data and extreme differences in the characteristics of water transport networks make it difficult to establish useful energy-intensity ranges for water transport at

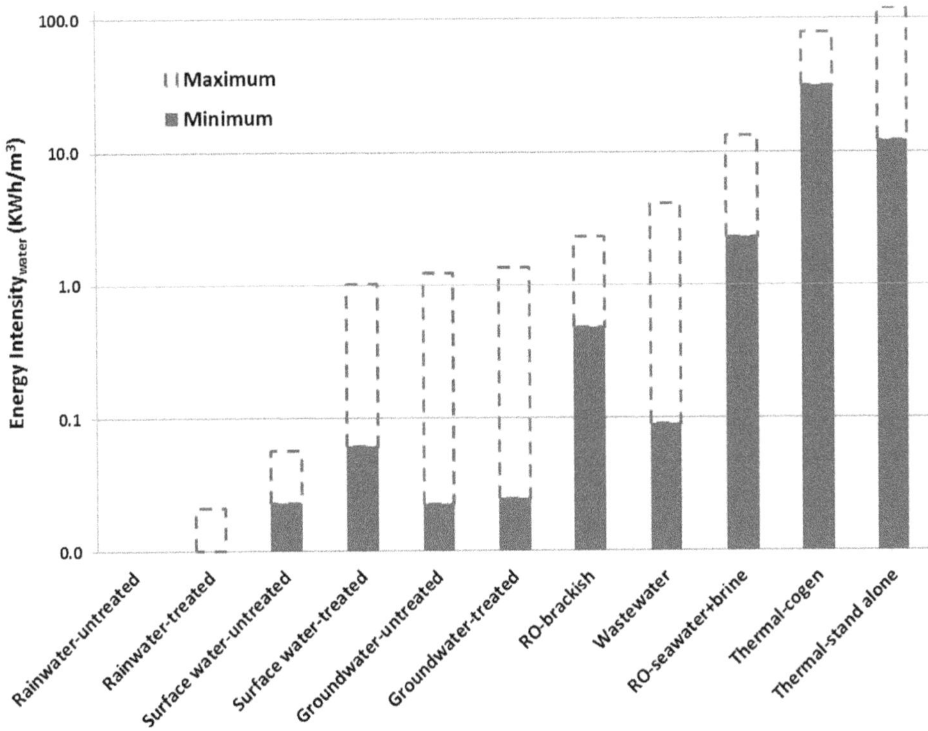

Figure 2. Energy required for different types of water. Sources: Al-Karaghouli and Kazmerski (2013), Abdel-Jawad (2001), Buenomena (2013), Darwish, Al-Najem, and Lior (2009), Drewes (2011), Global Water Intelligence (2015), Hamed (2004), Hayek (2014), Ludwig (2010), Matar, Murphy, Pierru, and Rioux (2014), Plappally and Lienhard (2012), Veera (2011), Vieira, Beal, Ghisi, and Stewart (2014). Note: Vapour compression (VC) desalination is a modified form of the MED process where the evaporation of seawater is obtained by the application of heat through a mechanical compressor. In this unique process, the energy requirements are lower than MSF/MED thermal cogeneration technologies, which is why the 'thermal stand-alone' technology has lower potential minimum energy requirements. In general, thermal cogeneration processes are less energy-intensive than thermal stand-alone processes.

the national level. For example, when water transport systems are gravity-fed, such as the All-American Canal, located in Southern California adjacent to the Mexican border, energy can be produced from hydroelectric power plants. The All-American Canal is used to transport 3.8 km^3 of water 82 miles from the Colorado River to the Imperial Valley (Imperial Irrigation District, 2015), and the process generates more than 253 million kWh of electricity on an annual basis. In contrast, the proposed Ebro project in Spain that sought to transfer 860 GL/y of water from the Ebro River a distance of 745 km to the south of the country was predicted to require roughly 4 kWh/m^3 (Marcuello, Capilla, Murillo, Barcones, & Meyer, 2003; Plappally & Lienhard, 2012). The Ebro water transfer project was cancelled in 2004 due to social, environmental and economic concerns. Studies suggested that the project would have led to the disappearance of the Ebro Delta, an important wetland in Spain, and that the economic costs would have been higher than initial government estimates (World Wide Fund for Nature [WWF], 2003).

According to Plappally and Lienhard (2012), energy intensity of water transport systems is location-specific, and dependent on factors such as pipeline grade, the soil's seepage or percolation properties, solar radiation and climatic behaviour in a given geographical region. Although difficult to generalize, energy for water transport can represent a significant proportion of the total energy required to meet water demand. For example, Hardy, Garrido, and Juana (2012) have estimated that 21% of the total energy used to meet water demand in Spain goes to distribution and water transport. It should be noted that while historically water transport was primarily used to address municipal water-scarcity challenges, it is increasingly being used for agriculture and industry.

Water scarcity and energy use

The water scarcity of a country is traditionally measured as the ratio of its water withdrawals to total renewable water resources (Brown & Matlock, 2011). If a country withdraws less than 20% of its total renewable water resources it is considered water-abundant. When a country withdraws between 20% and 40% of total renewable water resources it is considered water-scarce; and when the ratio exceeds 40%, the country is considered severely water-scarce.

Figure 3 estimates the relationship between water scarcity and the aggregate energy intensity required to provide a cubic metre of water in a sample of countries. Water-scarcity ratios were obtained from Aquastat (http://www.fao.org/nr/water/aquastat/data/query/) and national accounts. Energy data were compiled by estimating three sources: untreated surface and groundwater, treated surface and groundwater, and desalinated unconventional water. Given data limitations for numerous countries, energy for interbasin water transport and wastewater treatment were not included in the calculations, and so the energy intensity is probably underestimated, particularly in water-scarce regions where large interbasin water transfers occur or where substantial wastewater treatment occurs.

The energy required to extract untreated surface water was estimated to be 0.023 kWh/m^3 for OECD countries and 0.034 kWh/m^3 for non-OECD countries. These values were calculated based on a theoretical lift parameter (l, 2.73 kWh to lift 1000 m^3 a height of 1 m), a hydraulic head (h) of 5 m, and a pump efficiency (ε) of 40% for non-OECD countries and 60% for OECD countries, and described in the equation below.

$$e_{GW} = \frac{l * h}{\varepsilon}$$

This equation has been used to estimate energy for water extraction in a number of studies, notably that of Nelson et al. (2008). While the efficiencies of pumps can vary within and between countries, particularly when both electrical and diesel pumps are used, the assumptions and results obtained are in line with case studies from both developed and emerging economies (Abadia, Rocamora, Ruiz, & Puerto, 2008; Japikse, Marscher, & Furst, 1997; Moreno, Carrion, Planells, Ortega, & Tarjuelo, 2007; Shah, 2009). Energy for untreated groundwater was estimated in the same manner, with groundwater depths coming from a variety of government figures and academic studies. Energy for treated water in countries came from academic studies and reports from utilities, while energy for desalinated water was derived from numerous academic and government sources (see sources section of Figures 3 and 6). Despite the use of aggregated data, sensitivity analysis in which pump efficiencies and energy requirements for water treatment were varied under multiple scenarios did not greatly alter the aggregate relationships between countries. Given this, one can assume that the analysis above offers a reasonable estimate of the relationship between water scarcity and energy intensity for water in the sample of countries.

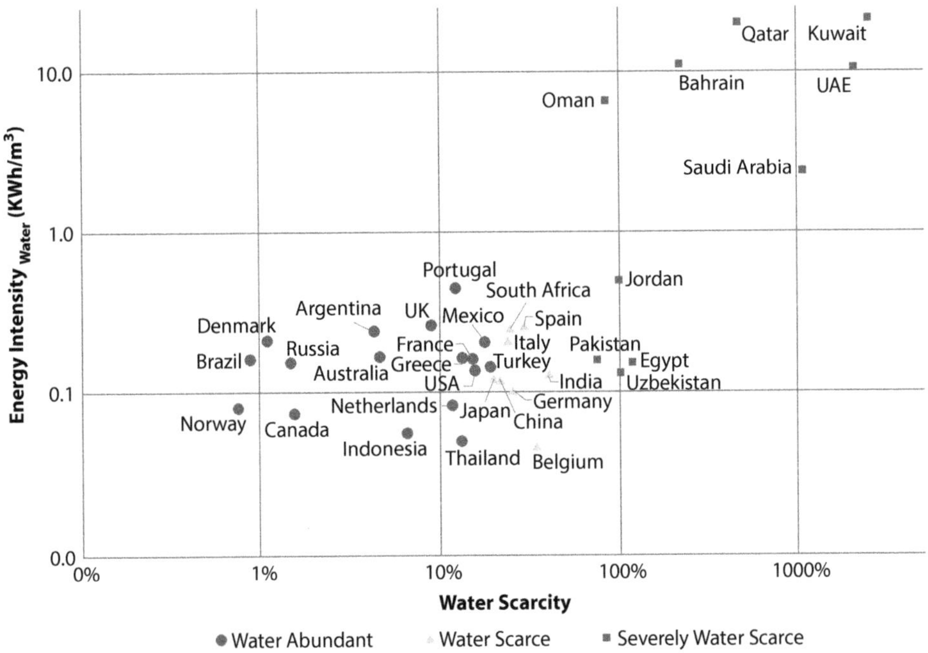

Figure 3. Energy intensity of water (extraction and treatment) and water scarcity in a sample of countries. Sources: Abderrahman (2001), Agenda 21 Della Terra d'Arneo (n.d), Al-Karaghouli & Kazmerski (2013), Al-Mashaikhi (2011), Al-Mooji, Hofstetter, & Renck (2013), Abdel-Jawad (2001), Aqualogy (http://www.aqualogy.net), Aquastat (http://www.fao.org/nr/water/aquastat/data/query/), Association of Private Water Operators (2013), Australian Bureau of Statistics (2014), Basharat (2012), British Geological Survey (2015), Buenomena (2013), Bundesanstalt für Geowissenschaften und Rohstoffe (2015), Burton (1996), Campanelli, Foladori, and Vaccari (2013), Chan (2013), China Urban Water Association (2012), Chudaeva et al. (2008), CONAGUA (2014), CONUEE (2011), Copeland (2014), Darwish et al. (2009), Department for Environmental, Food and Rural Affairs (2008), Department of Environment, Water and Natural Resources (2014), Dimakis, Colleuille, & Wong (2005), Drewes (2011), ECRA (2011), El Tahlawi, Farrag, and Ahmed (2008), Encyclopedia of Desalination and Water Resources (2014), Entidade Reguladora dos Serviços de Águas e Resíduos [ERSAR], (2012), Environmental Agency Abu Dhabi (2014), European Benchmarking Cooperation (2014), European Environment Agency (2014), Frijns, Mulder, and Roorda (2008), Gaut (2010), Global Water Intelligence (2015), Godskesen, Hauschild, Rygaard, Zambrano, and Albrechtsen (2013), Hadian, Mardiana, Abdurahman, and Iman (2006), Hamed (2004), Hardy et al. (2012), Hayek (2014), Hernández-Mora, Martinez-Corona, Llamas-Madurga, and Custodio-Gimena (2010), Imperial Irrigation District (2015), International Groundwater Resources Assessment Centre [IGRAC] (https://ggmn.unigrac.org), Japan Waterworks Association (2014), Lemos, Dias, Gabarrell, and Arroja (2013), Ludwig (2011), Karimov et al. (2015), Kenway et al. (2008), Kumar (2013), Kwanyuen, Mainuddin, and Cherdchanpipat (2003), Li, Liu, Zheng, Han, and Hoff (2015), Maas (2009), MacHarg and McClellan (2004), Matar et al. (2014), McLay (2005), McMahon and Price (2011), Margat and van der Gun (2013), Ministry of Water & Electricity (2010), Ministry of Water and Irrigation (2015), Nelson et al. (2008), Natural Resources Canada Database (2014), Papapetrou, Wieghaus, & Biercamp (2010), Pearce (2007), Peng (2014), Plappally and Lienhard (2012), Portela and Cohim (2011), Qatar General Electricity and Water Corporation (2014), Queensland Department of Natural Resources and Mines (2014), Raucher et al. (2010), Reboucas (1999), Renzoni and Germanin (2007), Saatçı (2015), Shah (2009), Tao (2012), Tech Archival (2014), Trans Adriatic Pipeline-TAP (2015), Shimizu, Dejima, and Toyosada (2012), Veera, Nirmalakhandan, and Deng (2010), Veera (2011), Venkatesh (2011), Vince (2007), Wang et al. (2012), Water in the West (2013), Water Resources Policy Division, Land and Water Bureau (2006), World Health Organization and UN ICEF (2014).

It is often assumed that energy intensity and water scarcity are positively correlated. As Figure 3 shows, this is true when the severely water-scarce countries of the Gulf region are considered. The Gulf countries contain almost no surface water, and so they rely on desalination and deep underground fossil aquifers in order to meet water demand. For example, the United Arab Emirates obtains 24% of its water from desalination, while 70% comes from fossil aquifers, and only 6% from surface water. Fossil aquifers take thousands of years to form, and when water is withdrawn from them, the total water available for future generations decreases. The Environment Agency of Abu Dhabi estimates that water levels in the agricultural region of Al Khazna fell from 56 m to 96 m between 1999 and January 2014, while water levels just north, in Sweihan, fell from 46 m to 104 m between 1998 and 2013 (Malek, 2015). The non-renewable nature of these aquifers causes them to fall when water is extracted; and as water levels drop, the energy costs for extraction rise.

When excluding the Gulf region, however, there is effectively no correlation between the energy intensity of water extraction and water scarcity. For example, the severely water-scarce countries of Pakistan and Egypt use less energy per unit of water extracted than Portugal, a water-abundant country. The primary differences between these countries are twofold. First, while Portugal enjoys abundant renewable freshwater resources, these resources exist as groundwater, which, as stated, requires more energy for extraction compared to surface water. In Portugal, 74% of the total water supply comes from underground aquifers. In contrast, Egypt obtains 89% of its water from the surface, primarily the Nile Delta, while Pakistan obtains 65% of its water from lakes and rivers. Thus, while the total renewable water resources in Egypt and Pakistan are being exhausted more quickly than in Portugal, the energy requirements for this extraction are lower, as the water is being extracted from more accessible sources.

Second, water scarcity estimated at the national level can be misleading. This is because regional differences within a country can have a profound impact on actual water resources readily available to the populace. Portugal is not considered a water-scarce country because of the abundant water resources in the country's north. Much of the country's population and agricultural activity, however, are located south of the city of Porto, where water resources are scarcer, and so groundwater from deeper sources must be withdrawn. In Egypt and Pakistan, while renewable water resources are scarcer at the national level, they are much more accessible to the population, and thus require less energy. The situation is similar in Australia, a country considered water-abundant due to the extensive water resources located in the tropical north, despite the fact that an estimated 65% of the population live in water-scarce regions (own calculations, data from Australian Bureau of Statistics, 2015; Chartres & Williams, 2006; *The Economist*, 2007; World Bank, 2015). While imperfect, national water-scarcity calculations are still useful as they offer a theoretical indication of the water resources that can be mobilized by the state for extraction and consumption.

An additional factor that can affect the relationship between water scarcity and energy intensity is total water withdrawals. Consider the cases of the United Kingdom and Spain. The two countries have similar energy intensities despite the fact that Spain is a more water-scarce country, uses far more desalination, has deeper average groundwater bores, and extracts more than twice the groundwater (roughly 5.7 km^3 compared to 2.16 km^3; http://www.fao.org/nr/water/aquastat/data/query/). The similarities in energy intensity are due to the way water is used in agriculture. In the UK, the total value added by agriculture in 2013 was only USD 12 billion, or 0.5% of the nation's total GDP (in constant 2005 USD; United

Nations Statistics Division, 2015). The UK enjoys ample rainfall, and so it only extracts roughly 1.3 km^3 of water to achieve that agriculture production. Spain is a far dryer country, but in 2013 its agriculture sector contributed 3% to the GDP, a figure representing USD 35 billion. Because of the sector's size and the country's climate, Spain must extract 20 km^3 of water for agriculture. While Spain extracts far more water for agriculture than the UK, over 79% of this water comes from the surface. This significantly reduces the *average energy intensity* required to meet water demand. This result, however, is somewhat misleading because Spain uses far more *total energy* to meet water demand given the higher overall levels of water extraction.

This section has assessed the supply of water resources. It has demonstrated that water comes from a number of different sources that have varying energy requirements. In theory, water-abundant countries will have lower energy intensities for water withdrawals. In practice, this relationship is complicated by factors such as regional water scarcity, the location of total renewable water resources, and total water use. The analysis also suggests that as countries exhaust easily obtainable water resources they are likely to move to more expensive, and energy-intensive, water resources. This phenomenon is prevalent in the Gulf region. It should be noted that energy-intensive water sources are, in general, also the most costly, and so as countries move up the marginal cost of energy curve, the total costs of water often increase (Figure 2).

The extent to which countries must shift to more energy-intensive water resources is a function of not only water supply but also how water is used by the population and the economy. The demand for water is addressed in the following section.

The determinants of water demand

Only a small percentage of water extracted is consumed *directly* by the population. Globally, roughly 70% of all water withdrawn is used for agriculture, while 19% is an input for industrial processes (http://www.fao.org/nr/water/aquastat/data/query/). Of the remaining 11% of water allocated to municipalities, much is used either by firms (local companies connected to the municipal grid) or for irrigation in residential communities. Given that so much extracted water is used as an input for production, it is appropriate to consider the resource from the perspective of production theory, whereby output (Y) will be some function of a combination of inputs, such as capital (K), labour (L), land (H) and water (W):

$$Y = F(K, L, H, W)$$

The relative contribution of each factor of production to output is determined by comparative productivity. Productivity is defined as output (typically represented in physical or monetary terms) per unit input (represented in terms of monetary costs or input units). In the short run, it is almost always possible to increase the productivity of an individual input. Without technological innovation, however, this increase will come at the expense of the productivity of other factors. For example, a farmer could reduce the amount of extracted water required to produce crops (i.e. increase water productivity) by investing in drip-irrigation technologies (i.e. increasing capital expenditures, and thus decreasing capital productivity) or employing more labour to irrigate plants directly by hand (i.e. increasing labour expenditures, and thus decreasing labour productivity). A farmer will only choose to substitute water for capital or labour if the cost savings from using less water are greater than the increased expenditures from employing more capital or labour. In cases where the costs of installing

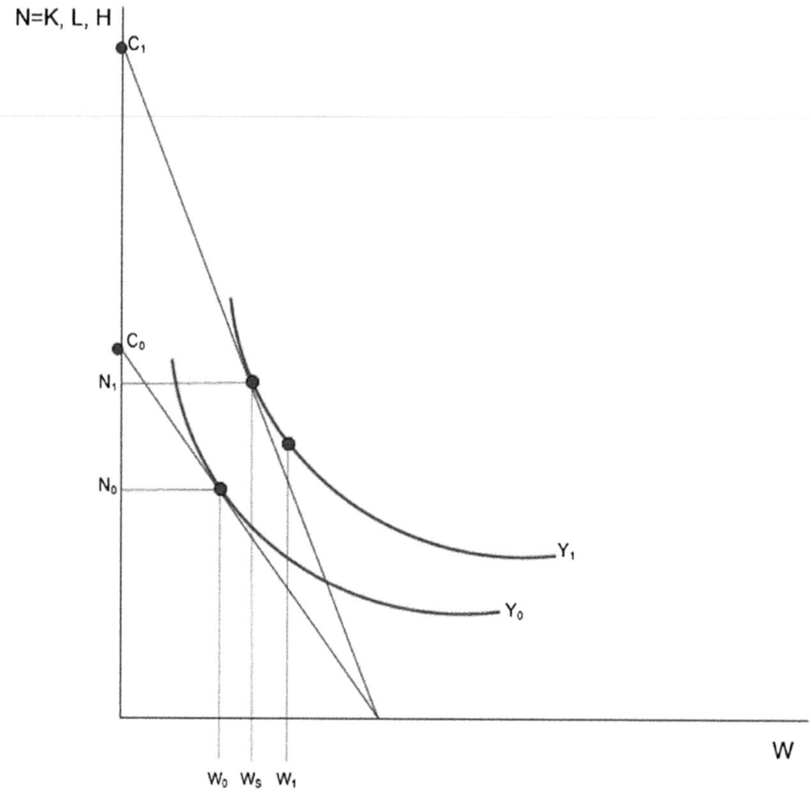

Figure 4. Potential effects of capital and labour subsidies on water consumption. Adapted from Sorrell (2014).

drip irrigation or increasing labour use are higher than the savings from using less water, the investment will not make economic sense, as it will increase the total costs of production. It should be noted that, because energy can represent up to 25% of groundwater system production costs (Dhuyvetter, O'Brien, Haag, & Holman, 2014; Zilberman, Sproul, Rajagopal, Sexton, & Hellegers, 2008), the cost of water use for agriculture may vary considerably with fluctuations in energy costs: the higher the energy costs, the more incentive there may be for a farmer to invest in water savings.

The amount of water used for production is a function of the relational costs between water and other factors of production. Water is used excessively in production when it is valued less than substitute inputs. Malik (2002) has argued that "inefficient use of both water and electricity in large part is also the result of non-metering of the consumption and the non-remunerative tariff structures for these services". The effects of this inefficiency are not restricted to producers, but can also affect how embodied water is consumed. Allan, Keulertz, and Woertz (2015) have noted that the true values of water and energy are not reflected in the prices of food and manufactured commodities paid by consumers in private-sector markets. As a result, the quantity demanded of these products, and thus the total water and energy used in their production, can rise.

If policy makers seek to change the relative contributions of each input to production, there are effectively two options: decrease the costs of substitute inputs (such as labour and capital) or increase the cost of water. The first option is achieved through subsidies. For example, farmers could be offered more efficient drip-irrigation technologies at subsidized prices, or be offered tax rebates for employing more labour. The second option is achieved through increasing the price of water, either by introducing a tax or increasing the prices of water inputs, namely energy, which, as stated, is often the largest single cost for water provision.

Both options have drawbacks. Figure 4 shows the potential effects of a capital or labour subsidy on water consumption. In the figure, vector *C* represents the total costs of agricultural production while isoquant *Y*, which is assumed to be homothetic, represents the possible mix of water *W* and the nest, *N*, of land, labour and capital, for production. In the scenario, prior to the introduction of the subsidy the point of tangency is $\{Y_0, C_0\}$. This intersection represents the chosen mix of factors of production given cost constraints. A subsidy on land, labour, and/or capital decreases the total costs of *N*, which leads to an outward shift of the total cost curve, from C_0 to C_1. This movement has two effects.

First, there is an overall increase in production, as cheaper inputs allow for increased production for the same costs. This increase in production leads to a shift in water use from W_0 to W_1. Second, there is a substitution of *W* for *N*, which leads to a movement upward along the *Y* isoquant, from W_1 to W_S, where S represents subsidy. The net effect of the subsidy on aggregate water consumption will depend on the relationship between the substitution and cost effects. In instances where the cost effect is greater than the substitution effect ($W_1 - W_0 > W_1 - W_S$), the subsidy will lead to an increase in aggregate water consumption, from W_0 to W_S, as demonstrated in Figure 4.

In practice, this effect could be described as follows. When labour or capital is subsidized, the total costs of production are reduced. While this may reduce water usage per unit of output, it may also increase the profitability of the industry, thus encouraging more production (either by incumbents or new entrants), and increasing aggregate water use. Also, if producers are operating in a competitive market, cost reductions may cause end prices for consumers to decrease, which can increase quantity demanded. The lack of aggregate savings from technologies meant to raise efficiency is known as Jevons' paradox (Alcott, 2005), or the rebound effect. Empirical studies suggest that irrigation modernization may lead to a rebound effect that negates efforts to improve water productivity. Pfeiffer and Lin (2014) found that irrigation efficiency in western Kansas from 1995 to 2005 led to increases in groundwater extraction of over 100%. The results were due primarily to switching to more profitable, but also more water-intensive, crops. Similarly, Ward and Pulido-Velázquez (2008) showed that water conservation subsidies in the Rio Grande Basin have not reduced aggregate water use, as farmers have simply increased yields. Because energy use is directly related to water use, when water consumption increases so does aggregate energy consumption.

The second option, increasing the price of water, offers a better economic solution for improving water productivity and decreasing energy consumption. As Figure 5 shows, a water tax leads to increased costs for water, causing an inward shift from C_0 to C_1. This shift also has two effects. First, lower production reduces total water use, from W_0 to W_1. Second, higher water costs make other factors more attractive, leading to a substitution effect equal to $W_1 - W_T$, where T represents tax. Thus, the total reduction of water resulting from the tax is $W_0 - W_T$.

In practice, the effect of a tax could be described as follows. An increase in the price of water will raise the total costs of production, which will result in reduced production and/

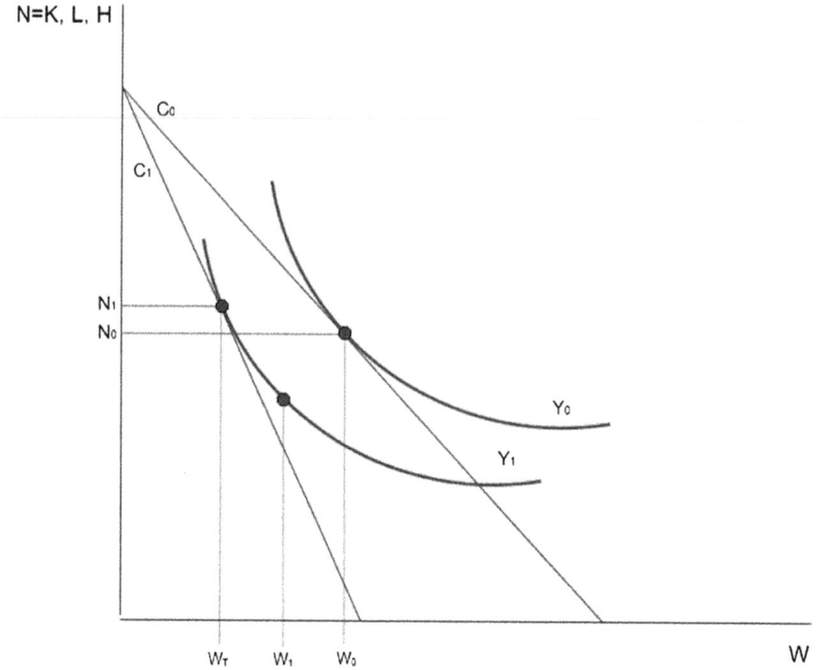

Figure 5. Potential effects of a water tax on water consumption. Adapted from Sorrell (2014).

or lower quantity demanded of products. In principle, this reduction in water use will lead to lower energy consumption for the provision of that water. While a tax may be an economically favourable solution, it is rarely implemented, as it is politically sensitive given the reductions it would entail in firm profitability as well as consumer welfare.

It should be noted that the logic above also applies to households using municipal water supplies. Households could be incentivized to reduce water use (for example by adopting low-flow faucets and showerheads) either by increasing the price of water or by subsidizing the prices of water-conserving technologies. The effects of household water reductions on energy are significant. This is because the energy savings related to the reduction of water use by household items such as washing machines, dishwashers and showers, although often overlooked, can be as high as 11.2 times the energy used to deliver water services (Kenway et al., 2008). As with water use for production, however, subsidies can lead to a rebound effect (i.e. increases in aggregate water use), while taxes can lead to cost increases (i.e. a decrease in welfare). It should be noted that even though taxes may be preferable, finding the correct level of tax can be difficult. Easter and Liu (2007) have suggested that one reason private and social costs of water are not included in tariffs is that they are difficult to estimate and may be beyond the knowledge of many countries. The authors argue for additional reforms to complement increases in water tariffs. For example, volumetric pricing, as opposed to simply charging farmers per hectare irrigated, could improve the efficiency of water used. Likewise, water quotas can be used to force water use reductions.

The analysis above suggests that while economic solutions for managing water resources exist, they can be politically sensitive, and therefore difficult to implement. This often results in overuse of water by households, industry and agriculture. Overuse has two implications for energy. First, it increases overall energy consumption, as energy use is a function of total water withdrawn. Second, and perhaps more importantly, high water use can necessitate a movement to more energy-intensive water sources, thus increasing the average energy required to meet water demand. This is exemplified in Figure 6. The figure shows the per capita energy required to meet water demand in a sample of countries under two scenarios. The first scenario is a baseline, showing per capita energy for water under current consumption practices. The second scenario shows how this per capita energy use could change if the three most water-consuming crops in each country were removed from production. The top of the dotted bar (referred to as the Crop Reduction Scenario) assumes a reduction in the crop's actual water source, while the top of the solid bar (referred to as the Crop Reduction + Zero Desalination Scenario) assumes that water reductions occur in the most costly and energy-intensive sources, such as desalination and deep groundwater withdrawals. Any level within the dotted area represents some combination of a reduction in actual water sources and the most energy-intensive water sources.

Figure 6 suggests that crop elimination has the potential to dramatically reduce energy intensity for water extraction in some countries. For example, in 2010 Saudi Arabia used 2301 kWh per capita to extract and treat water for municipalities, industry and agriculture. If the country's three most water-intensive crops – alfalfa, dates and wheat – were eliminated in that year, total freshwater withdrawals would have been reduced from 25 km^3 to 15 km^3. If reductions occurred from each crop's actual source (typically deep groundwater aquifers), per capita energy consumption for water would have been reduced to 2094 kWh. Saudi Arabia, however, desalinated roughly 1.3 km^3 of water in 2010. Eliminating the three crops would reduce the need for this expensive and energy-intensive desalination, as the desalinated water could be replaced in part or in full by conventional water, effectively allowing the country to move down the marginal cost of energy curve. If all of the country's desalinated water were replaced by conventional groundwater, this would reduce the average energy required to meet water demand by 87%. It should be noted that the water reductions from eliminating these crops would decrease total freshwater consumption by 41%, which would still result in Saudi Arabia being a severely water-scarce country, but non-renewable water resources would be depleted at a much lower pace. The situation is similar in Oman. Eliminating the three most water-intensive crops in 2010 would have significantly reduced the average energy intensity of water extraction, as well as reducing water scarcity by 53%. As with a water tax, eliminating crops is a politically sensitive topic, as some countries consider this a surrendering of food security. Despite this, recent developments suggest that it is possible. The government of Saudi Arabia has mandated the elimination of wheat production by 2016, and Almarai, the country's largest dairy producer, is employing a strategy to import all of the alfalfa it uses as fodder (Almarai, 2014). In 2008, Almarai used roughly 400,000 tons of domestically produced alfalfa as fodder (Almarai, 2009). This figure represented 20% of the total fodder used by the company that year, and roughly 16% of the total alfalfa that was produced in Saudi Arabia. Eliminating this alfalfa would reduce water consumption by 0.9 km^3. If similar commitments were made by other agriculture producers that rely on alfalfa as feed, and domestic alfalfa production were eliminated, this would reduce water consumption by an estimated 6 km^3 (FAOStat, http://faostat3.fao.org/home/E; Ministry of Agriculture, 2014; Multsch, Al-Rumaikhani, Frede, & Breuer, 2013). While dates are more politically and culturally sensitive, the elimination of both fodder and wheat would reduce water consumption by 7.3 km^3 (FAOStat, 2015; Ministry of Agriculture, 2014; Multsch et al., 2013), which is far more than the current production of desalinated water. These types of

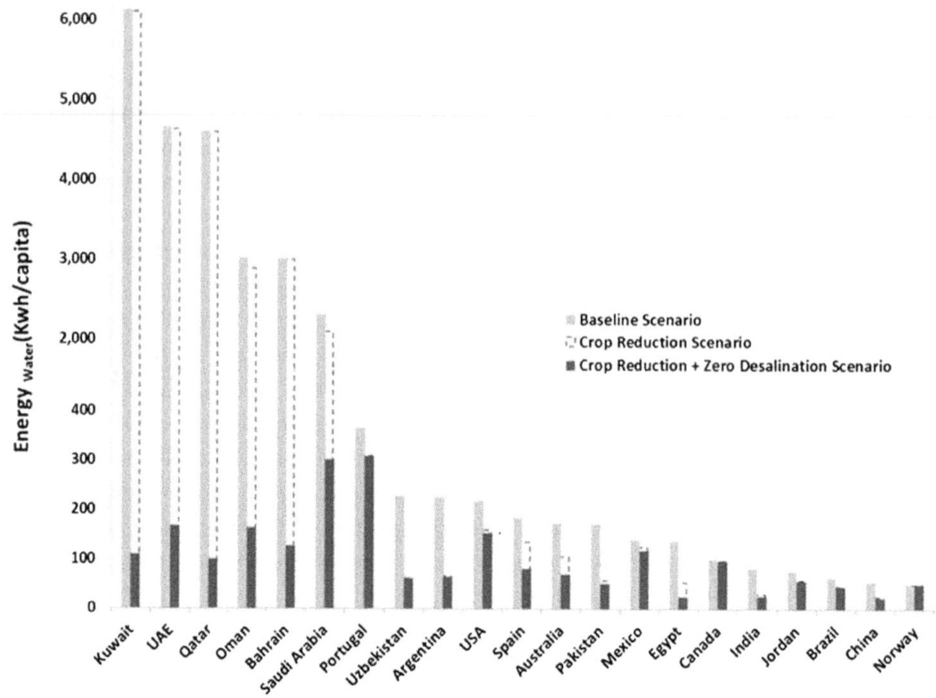

Figure 6. Energy for water per capita, baseline and crop elimination. Sources: Abderrahman (2001), Agenda 21 Della Terra d'Arneo (n.d), Al-Karaghouli & Kazmerski (2013), Al-Mashaikhi (2011), Al-Mooji, Hofstetter, & Renck (2013), Abdel-Jawad (2001), Aqualogy (http://www.aqualogy.net), Aquastat (http://www.fao.org/nr/water/aquastat/data/query/), Association of Private Water Operators (n.d), Australian Bureau of Statistics (2014), Basharat (2012), British Geological Survey (2015), Buenomena (2013), Bundesanstalt für Geowissenschaften und Rohstoffe (2015), Burton (1996), Campanelli et al. (2013), Chan (2013), Chudaeva et al. (2008), CONAGUA (2014), CONUEE (2011), Copeland (2014), Darwish et al. (2009), Department for Environmental, Food and Rural Affairs (2008), Department of Environment, Water and Natural Resources (2014), Dimakis, Colleuille, & Wong (2005), Drewes (2011), ECRA (2011), El Tahlawi et al. (2008), Encyclopedia of Desalination and Water Resources (2014), Environmental Agency Abu Dhabi (2014), European Benchmarking Cooperation (2014), European Environment Agency (2014), Frijns et al. (2008), Gaut (2010), Global Water Intelligence (2015), Godskesen et al. (2013), Hadian et al. (2006), Hamed (2004), Hardy et al. (2012), Hayek (2014), Hernández-Mora et al. (2010), Imperial Irrigation District (2015), International Groundwater Resources Assessment Centre [IGRAC] (https://ggmn.unigrac.org), Japan Waterworks Association (2014), Lemos et al. (2013), Ludwig (2011), Karimov et al. (2015), Kenway et al. (2008), Kumar (2013), Kwanyuen et al. (2003), Li et al. (2015), Maas (2009), MacHarg and McClellan (2004), Matar et al. (2014), McLay (2005), McMahon and Price (2011), Margat and van der Gun (2013), Ministry of Water & Electricity (2010), Nelson et al. (2008), Natural Resources Canada Database (2014), Papapetrou, Wieghaus, & Biercamp (2010), Pearce (2007), Peng (2014), Plappally and Lienhard (2012), Portela and Cohim (2011), Qatar General Electricity and Water Corporation (2014), Queensland Department of Natural Resources and Mines (2014), Raucher et al. (2010), Reboucas (1999), Renzoni and Germanin (2007), Saatçı (2015), Shah (2009), Tao (2012), Tech Archival (2014), Trans Adriatic Pipeline-TAP (2015), Shimizu et al. (2012), Veera et al. (2010), Veera (2011), Venkatesh (2011), Vince (2007), Wang et al. (2012), Water in the West (2013), Water Resources Policy Division, Land and Water Bureau (2006), World Health Organization and UNICEF (2014). Note: Energy values from KAPSARC analysis on energy for surface water abstraction, groundwater abstraction and desalination. Crop water use data from the Water Footprint Network (average values between 1996 and 2005). Crop production data from FAOSTAT for the year 2011 (http://faostat3.fao.org/home/E).

initiatives are necessary to help the country reach its goal of reducing water withdrawals by 30% by 2030, and if coupled with reductions in energy-intensive water supplies, could also lead to energy reductions.

In addition to estimating the potential benefits from crop elimination in some countries, there are two other important findings from the figure. First, in the four other Gulf Cooperation Council (GCC) countries, Kuwait, the United Arab Emirates, Qatar and Bahrain, eliminating water-intensive crops has effectively no impact on average energy used for water, as no water-intensive crops are grown on a large scale in these countries. For energy intensity reductions to occur, crop elimination must be coupled with a movement away from desalination towards conventional water. This strategy is unfavourable, however, because a shift away from desalinated water would increase aggregate groundwater use, which would raise the overall water scarcity of the countries. For these severely water-scarce countries, where desalination is required, two options exist for reducing energy use: reducing municipal water consumption; and/or adopting more energy-efficient desalination technologies (i.e. a movement from thermal technologies to membrane technologies). Second, some countries, like Mexico, Canada and Norway, would enjoy only minimal reductions in average energy consumption with the elimination of water-intensive crops. This is because most crops in these countries rely on rainwater, and so energy use for extracted water is marginal. For example, over 99% of the water used to grow wheat in Canada and Norway comes from rainfall (own calculations, data from Mekonnen & Hoekstra, 2011). In cases where crop elimination does not affect energy intensity, countries may consider reducing municipal water consumption, or adopting more efficient pumps for extraction.

Virtual trade in water: implications for water and energy savings

Allan (1998) has defined the virtual-water content of a commodity, good or service as the volume of water used in its production. The trade in virtual water, therefore, represents the amount of water embedded in products traded internationally. Using this framework, Allan's work has explored how food trade may affect the water economies and policies in water-scarce countries (Allan, 2003). He has argued that importing water-intensive crops offers a policy option for countries with water deficits, which may even be particularly valuable for preventing 'hydropolitics' from sparking full-scale conflicts (Allan, 2002).

Building on Allan's insights, Hoekstra and Hung (2002) examined the volume of virtual water trade flows between countries, putting national water trade balances in the context of domestic water requirements and resources. This work offers an empirical assessment of the way countries are water-dependent on others. More recent work in this area has suggested that "about one-fifth of both global cropland and agricultural water use is allocated to the production of agricultural commodities consumed abroad" (Hoekstra & Mekonnen, 2012, quoted in MacDonald et al., 2015).

The high potential for virtual water trade to improve the sustainability of water resources in water-scarce countries should be complemented with an analysis of energy savings. As shown in Figure 6, crop elimination, which is effectively a substitution away from domestic production to imports, could lead to significant reductions in both total energy use and, in some cases, the intensity of energy used to meet water demand, both of which would have positive effects on sustainability and the environment.

Conclusion

This article offers a framework for evaluating how energy is used to meet water demand. The supply of and demand for water, and the energy required to withdraw that water, were disaggregated. The supply analysis demonstrates that water comes from a number of different sources that have varying energy requirements. In theory, water-abundant countries will have lower energy intensities for water withdrawals. In practice, this relationship is complicated by factors such as regional water scarcity, the location of total renewable water resources, and total water use. The analysis also suggests that as countries exhaust easily obtainable water resources they are likely to move to more expensive, and energy-intensive, water resources. This phenomenon is prevalent in the Gulf region as well as water-scarce regions in developed countries like Spain, Australia and parts of the United States.

Regarding demand, while economic solutions for managing water resources exist, such as raising water prices or mandating the elimination of some agriculture crops, they can be politically sensitive, and therefore difficult to implement. It should be noted that some countries have made progress in these areas. For example, in the mid-1970s, Denmark experienced two years of drought, in which city water supplies were virtually exhausted. This resulted in campaigns to conserve water as well as the introduction of water taxes, which led to a decrease in aggregate municipal water consumption from 605 million m^3 in 1980 to 400 million m^3 in 2005 (Geological Survey of Denmark & Greenland, [GEUS], 2006). In addition, as stated, Saudi Arabia has reduced its wheat production since 2008, and plans to rely solely on imported wheat by 2016. The decision was made because of the negative impact domestic wheat production has on water resources. The country is trying to meet its food security goals through other means, such as the creation of the Saudi Agricultural and Livestock Investment Co. (SALIC), which has been endowed with USD 800 million to improve food security in the country through foreign purchases. In 2013, SALIC purchased 40,000 hectares of active farmland in Poland and the Ukraine; and in 2015 the group became a majority investor in the Canadian Wheat Board. Should these initiatives lower agricultural water use in the country, there may be potential reductions in the aggregate energy used for water, as well as average energy intensity for water withdrawals.

The article has suggested that while supply and demand management can reduce the energy required for water resources, uncontrollable factors may limit the potential efficacy of certain policy options. For example, in severely water-scarce regions, it is likely that desalinated and/or deep aquifer water are necessary components of the water portfolio, and so these regions will always have a higher overall energy intensity than countries with abundant, clean surface water. Similarly, it is difficult for water-abundant countries that already rely on rainwater and surface water to make significant improvements to energy intensity. In both cases, focus should be placed on reducing water consumption, as this will lower overall energy expenditures for water.

Last, it should be noted that the analysis in this article is based on aggregated data from multiple sources. In some countries, water and energy data are granular and current, while in others water resources and energy use are based on estimates. Moving forward, it behoves countries to improve the way water and energy data are collected and published so that more cross-country research can be done. It is hoped that this special issue contributes, in part, to this initiative.

Acknowledgement

The authors would like to acknowledge the research assistance of Mathieu Carey, intern, King Abdullah Petroleum Studies and Research Center.

Disclosure statement

No potential conflict of interest was reported by the authors.

References

Abadia, R., Rocamora, C., Ruiz, A., & Puerto, H. (2008). Energy efficiency irrigation distribution networks I: Theory. *Biosystems Engineering, 101*, 21–27. doi:10.1016/j.biosystemseng.2008.05.013.

Abdel-Jawad, M. (2001). *Energy sources for coupling with desalination plants in the GCC countries*. ESCWA Consultancy Report.

Abderrahman, W. (2001). Energy and water in arid countries: Saudi Arabia, a case study. *International Journal of Water Resources Development, 17*, 247–255. doi:10.1080/07900620120031306.

Agenda 21 Della Terra d'Arneo. (n.d.) *Analisi di qualita' delle Acque* [Quality analysis of water]. Retrieved from http://www.a21arneo.altervista.org/RapportoStatoAmbiente1/Index213.htm.

Alcott, B. (2005). Jevons' paradox. *Ecological Economics, 54*, 9–21. doi:10.1016/j.ecolecon.2005.03.020.

Al-Karaghouli, A. & Kazmerski, L. L. (2013). Energy consumption and water production cost of conventional and renewable-energy-powered desalination processes. *Renewable and Sustainable Energy Reviews, 24*, 343–356. doi:10.1016/j.rser.2012.12.064.

Allan, J.A. (1998). Virtual water: A strategic resource global solutions to regional deficits. *Ground Water, 36*, 545–546. doi:10.1111/j.1745-6584.1998.tb02825.

Allan, J. A. (2002). Hydro-peace in the middle east: Why no water wars? A case study of the Jordan River Basin. *SAIS Review, 22*, 255–272. doi:10.1353/sais.2002.0027.

Allan, J. A. (2003). Virtual water - the water, food, and trade Nexus. Useful concept or misleading metaphor? *Water International, 28*, 106–113. doi:10.1080/02508060.2003.9724812.

Allan, J.A., Keulertz, M., & Woertz, E. (2015). The water–food–energy nexus: an introduction to nexus concepts and some conceptual and operational problems. *International Journal of Water Resources Development, 31*, 301–311. doi:10.1080/07900627.2015.1029118.

Almarai. (2009). *Farming in the desert*. Retrieved from Organic Dairy and Pastoral website http://www.organicpastoral.co.nz/site/organic/images/Desert.pdf.

Almarai. (2014). *Further announcement from Almarai Company regarding working progressively towards the importation of 100% of its agricultural feed requirements* (Press Release 24/09/2014). Retrieved from Almarai website https://www.almarai.com/wp-content/uploads/2014/12/Press_Release_20140924_Feed_Supply_Update-Sep-2014_EN.pdf.

Al-Mashaikhi, K. S. A. (2011). *Evaluation of groundwater recharge in Najd aquifers using hydraulics, hydrochemical and isotope evidences (Unpublished doctoral dissertation)*. Jena, Germany: Faculty of Chemistry and GeosciencesFriedrich-Schiller-Universität.

Al-Mooji, Y., Hofstetter, E., & Renck, A. (2013). *Inventory of shared water resources in Western Asia*. Retrieved from Inventory of Shared Water Resources website: http://www.escwa.org.lb/information/publications/edit/upload/E_ESCWA_SDPD_13_Inventory_E.pdf .

Association of Private Water Operators. (2013). *Energy efficiency in urban water sector*. Retrieved from APWO website http://www.apwouganda.org/uganda-group-report-on-energy-efficiency-in-urban-water-sector-berlin-germany

Australia's water shortage: The big dry. (2007, April Article History 26). *The Economist*. Retrieved from http://www.economist.com/node/9071007.

Australian Bureau of Statistics. (2014). Water statistics. Retrieved from http://www.abs.gov.au/AUSSTATS.

Australian Bureau of Statistics. (2015). Australian demographic statistics. Retrieved from http://www.abs.gov.au/ausstats/abs@.nsf/mf/3101.0.

Basharat, M. (2012). Spatial and temporal appraisal of groundwater depth and quality in LBDC command – issues and options. *Pakistan Journal of Engineering and Applied Sciences, 11*, 14–29.

British Geological Survey. (2015). Groundwater levels timeline. Retrieved from http://mapapps.bgs.ac.uk/groundwatertimeline/home.html.

Brown, A. & Matlock, M. D. (2011). *A Review of water scarcity indices and methodologies (White Paper #106)*. Retrieved from The Sustainability Consortium website http://www.sustainabilityconsortium.org/wp-content/themes/sustainability/assets/pdf/whitepapers/2011_Brown_Matlock_Water-Availability-Assessment-Indices-and-Methodologies-Lit-Review.pdf.

Buenomena, G. M. (2013). Membrane processes for a sustainable industrial growth. *RSC Advances., 17*, 5694–5740. doi:10.1039/C2RA22580H.

Bundesanstalt für Geowissenschaften und Rohstoffe. (2015). Groundwater resources in Germany. Retrieved from http://www.bgr.bund.de/EN/Themen/Wasser/grundwasser_deutschland_en.html?nn=1548136.

Burton, F. L. (1996). *Water and wastewater industries. Characteristics and energy management opportunities* (Report No. CR-106941). Retrieved from the EPRI website http://www.epri.com/abstracts/Pages/ProductAbstract.aspx?ProductId=CR-106941

Campanelli, M., Foladori, P., & Vaccari, M. (2013). *Consumi elettrici ed efficienza energetica del trattamento delle acque reflue* [Electrical consumption and energy efficiency of wastewater treatment]. Santarcangelo: Maggioli editore.

Chan, A. (2013). *Characterising the water-energy nexus of in urban water system (Unpublished Master's thesis)*. Trondheim: Norwegian University of Science and Technology.

Chartres, C. & Williams, J. (2006). Can Australia overcome its water scarcity problems? *Journal of Developments in Sustainable Agriculture, 1*, 17–24.

China Urban Water Association. (2012). *China Urban Water Supply Yearbook*. Beijing: China Urban Water Association Press.

Chudaeva, V. A., Chudaev, O. V., Yurchenko, S. G., Sugimory, K., Matsuo, M., & Kuno, A. (2008). The composition of groundwater of Muraviov-Amursky Peninsula, Primorye. *Russia. Indian Journal of Geo-Marine Sciences, 37*, 193–199.

CONAGUA (Comisión Nacional del Agua). (2014). Estadisticas. Retrieved from www.conagua.gob.mx.

CONUEE (Comisión Nacional para el Uso Eficiente de la Energía). (2011). *Estudio de Bombeos Agropecuarios en Mexico* [Study on water pumping for agriculture and livestock in Mexico]. Retrieved from CONUEE website http://www.conuee.gob.mx/work/sites/CONAE/resources/LocalContent/7548/2/Informe_bombeo_AgricolaVF.pdf.

Copeland, C. (2014). *Energy-water nexus: The water sector's energy use*. Retrieved from Congressional Research Service website http://aquadoc.typepad.com/files/crs_energy_water_nexus_water_sectors_energy_use.pdf.

Darwish, M. A., Al-Najem, N. M., & Lior, N. (2009). Towards sustainable energy in seawater desalting in the Gulf area. *Desalination., 235*, 58–87. doi:10.1016/j.desal.2008.07.005.

Department for Environment, Food and Rural Affairs, UK. (2008). *Future Water: The Government's water strategy for England* (Policy Paper). Retrieved from the UK Federal Government website https://www.gov.uk/government/uploads/system/uploads/attachment_data/file/69346/pb13562-future-water-080204.pdf.

Department of Environment, Water and Natural Resources. (2014). *Water statistics in Australia*. Retrieved from the Government of South Australia website http://www.environment.sa.gov.au/Home.

Dhuyvetter, K. C., O'Brien, D. M., Haag, L., & Holman, J. (2014). *Center-pivot-irrigated corn: Cost-return budget in Western Kansas*. Retrieved from the Kansas State University bookstore website: http://www.ksre.ksu.edu/bookstore/pubs/mf585.pdf.

Dimakis, P., Colleuille, H., & Wong, W.K. (2005). *Pollution Impacts on Norwegian Groundwater Bodies*. Retrieved from Norwegian Water Resources and Energy Directorate website http://www.nve.no/Global/Publikasjoner/Publikasjoner%202005/Report%202005/report3-05.pdf.

Drewes, J. E. (2011). *An integrated framework for treatment and management of produced water* (Report No. 07122-12). Retrieved from the RPSEA website http://www.rpsea.org/files/1842/.

Easter, K. W. & Liu, Y. (2007). Who pays for irrigation: cost recovery and water pricing? *Water Policy, 9*, 285–303. doi:10.2166/wp.2007.006.

El Tahlawi, M. R., Farrag, A. A., & Ahmed, S. S. (2008). Groundwater of Egypt: An environmental overview. *Environmental Geology, 55*, 639–652. doi:10.1007/s00254-007-1014-1.

ECRA (Electricity & Cogeneration Regulatory Authority). (2011). *Water Situation Mastersheet* [Data set]. Retrieved January 12, 2015, from ECRA personnel.

Encyclopedia of Desalination and Water Resources. (2014). *Energy requirements of desalination processes*. Retrieved from http://www.desware.net/desa4.aspx.

Entidade Reguladora dos Serviços de Águas e Resíduos [ERSAR]. (2012). Energy and water in the public water supply and waste water sector. Retrieved from http://www.ersar.pt/.

Environmental Agency Abu Dhabi. (2014). Water resources. Retrieved from https://www.ead.ae/sitepages/home.aspx.

European Benchmarking Cooperation. (2014). Drinking water treatment. Retrieved from https://www.waterbenchmark.org/handlers/ballroom.ashx?function=download&id=38&rnd=0.22237726231105626.

European Environment Agency. (2014). Water base-groundwater: Characteristics of groundwater. Retrieved from http://www.eea.europa.eu/data-and-maps/data/waterbase-groundwater-10#tab-european-data.

Frijns, J., Mulder, M., & Roorda, J. (2008). *Op weg naar een klimaatneutrale waterketen* [The road to a climate neutral water chain] (Report No. A307729). Retrieved from Stowa website: http://waterenenergie.stowa.nl/Upload/publicaties/rapport%202008%2017.pdf.

Gaut, S. (2010). *Fractured aquifers: Evaluation of the drinking water source as a hygienic barrier*. Retrieved from Danva website http://danva.dk/Admin/Public/Download.aspx?file=Files%2FFiler%2F20+Sylvi+Gaut.pdf.

Geological Survey of Denmark and Greenland. (2006). *Water Supply in Denmark*. Retrieved from the Geological Survey of Denmark and Greenland website http://www.geus.dk/program-areas/water/denmark/vandforsyning_artikel.pdf.

Gleick, P. H. (Ed.) (1993). *Water in Crisis: A guide to the world's fresh water resources*. New York, NY: Oxford University.

Global Water Intelligence. (2015). *Desal database* [Desalination plants statistics]. Retrieved January 5, 2015, from www.desaldata.com.

Godskesen, B., Hauschild, M., Rygaard, M., Zambrano, K., & Albrechtsen, H.J. (2013). Life cycle and freshwater withdrawal impact assessment of water supply technologies. *Water Research, 47*, 2363–2374. doi:10.1016/j.watres.2013.02.005.

Hadian, M. S. D., Mardiana, U., Abdurahman, O., & Iman, M. I. (2006). Sebaran akuifer dan pola aliran air tanah di Kecamatan Batuceper dan Kecamatan Benda kota Tangerang, Propinsi Banten [The aquifer spreading and the groundwater pattern in Kecamatan Batuceper and Kecamatan Benda villages in Tangerang, province of Banten]. *Jurnal Geologi Indonesia, 1*, 115–128. doi:10.17014/ijog.vol1no3.20061.

Hamed, O. (2004). Overview of hybrid desalination systems: Current status and future prospects. *Desalination, 186*, 207–214. doi:10.1016/j.desal.2005.03.095.

Hardy, L., Garrido, A., & Juana, L. (2012). Evaluation of Spain's water-energy nexus. *International Journal of Water Resources Development, 28*, 151–170. doi:10.1080/07900627.2012.642240.

Hayek, B. (2014). *Energy efficiency in water pumping in Jordan*. Unpublished Manuscript.

Hernández-Mora, N., Martinez-Corona, L., Llamas-Madurga, M. R., & Custodio-Gimena, E. (2010). *Groundwater in the Southern Member States of the European Union: An assessment of current knowledge and future prospects in Spain*. Retrieved from EASAC website http://www.easac.eu/fileadmin/PDF_s/reports_statements/Spain_Groundwater_country_report.pdf.

Hoekstra, A. Y., & Hung, P. Q. (2002). *Virtual water trade: a quantification of virtual water flows between nations in relation to international crop trade*. (Value of Water Research Report Series No. 11). Retrieved from Water Footprint Network website http://waterfootprint.org/media/downloads/Report11.pdf.

Imperial Irrigation District. (2015). All-American Canal. Retrieved from http://www.iid.com/water/water-transportation-system/colorado-river-facilities/all-american-canal.

Japan Waterworks Association. (2014). *Water Supply in Japan 2013*. Retrieved from Japan Waterworks Association website: http://www.jwwa.or.jp/english/en_02.html.

Japikse, D., Marscher, W. D., & Furst, R. B. (1997). *Centrifugal pump design and performance*. Vermont: Concepts ETI Inc.

Karimov, A., Smakhtin, V., Khodjiev, K., Yakubov, S., Platonov, A., Karimov, A. A., & Avliyakulov, M. (2015). *Reducing the energy used to deliver irrigation water in Central Asia: Considering groundwater and managed aquifer recharge*. Unpublished Manuscript.

Kenway, S. J., Priestley, A., Cook, S., Seo, S., Inman, M., Gregory, A., & Hall, M. (2008). *Energy use in the provision and consumption of urban water in Australia and New Zealand* (Water for a Healthy Country Flagship Report Series). CSIRO and the Water Services Association of Australia. Retrieved from the CSIRO website http://www.clw.csiro.au/publications/waterforahealthycountry/2008/wfhc-urban-water-energy.pdf.

Kumar, P. (2013). *Energy efficiency opportunities and challenges in water supply System*. Retrieved from Alliance to Save Energy website: http://www.cseindia.org/userfiles/ASE_Opportunities%20and%20Challenges%20in%20Water%20Supply%20System_12August%202013_CSE_Puduchery.pdf.

Kumar, M. D., Scott, C. A., & Singh, O. P. (2013). Can India raise agricultural productivity while reducing groundwater and energy use? *International Journal of Water Resources Development, 29*, 557–573. doi:10.1080/07900627.2012.743957.

Kwanyuen, B., Mainuddin, M., & Cherdchanpipat, N. (2003). *Socio-ecology of groundwater Irrigation in Thailand*. Unpublished Manuscript.

Lemos, D., Dias, A. C., Gabarrell, X., & Arroja, L. (2013). Environmental assessment of an urban water system. *Journal of Cleaner Production, 54*, 157–165. doi:10.1016/j.jclepro.2013.04.029.

Li, X., Liu, J., Zheng, C., Han, G., & Hoff, H. (2015). *Energy for water utilization in china and options for policy reform*. Unpublished Manuscript.

Ludwig, H. (2011). Energy consumption of reverse osmosis seawater desalination: Possibilities for its optimization in design and operation of SWRO Plants. *Desalination and Water Treatment, 13*, 13–25. doi:10.5004/dwt.2010.982.

Maas, C. (2009). *Greenhouse gas and energy co-benefits of water conservation* (Report No. 09-01) Retrieved from Polis website http://poliswaterproject.org/sites/default/files/maas_ghg_.pdf.

MacDonald, G. K., Brauman, K. A., Sun, S., Carlson, K. M., Cassidy, E. S., Gerber, J. S., & West, P. C. (2015). Rethinking Agricultural Trade Relationships in an Era of Globalization. *BioScience, 65*, 275–289. doi:10.1093/biosci/biu225.

MacHarg, J. & McClellan, S. A. (2004). *Pressure exchanger helps reduce energy costs in brackish water RO system*. Retrieved from Ocean Pacific Technologies website http://www.ocean-pacific-tec.com/imagenes/news/11%20AWWA%20Journal%20Ocean%20Reef%20Brackish%20PX%2011-04.pdf.

Malek, C. (2015, February 21). Groundwater at danger level in UAE. *The National*. Retrieved from http://www.thenational.ae/uae/groundwater-at-danger-level-in-uae.

Malik, R. P. S. (2002). Water-energy nexus in resource-poor economies: The Indian experience. *International Journal of Water Resources Development, 18*, 47–58. doi:10.1080/07900620220121648.

Marcuello, J., Capilla, A., Murillo, J., Barcones, J., & Meyer, A. (2003). *Alegaciones al Proyecto de Transferencias Autorizadas por la Ley del Plan Hidrológico Nacional y Estudio de Impacto Ambiental* [Depositions of the Draft on Authorized Transfers by the National Hydrological Planning Law and Environmental Impact Assessment] (Report No. 03/03). Retrieved from Citarea website: http://citarea.cita-aragon.es/citarea/bitstream/10532/160/1/10532-80_136.pdf

Margat, J. & van der Gun, J. (2013). *Groundwater around the world: A geographic synopsis*. London: Taylor & Francis Group.

Matar, W., Murphy, F., Pierru, A., & Rioux, B. (2014). *KAPSARC Energy Model [Partial equilibrium model formulated as a mixed complementarity problem]*. Riyadh: KAPSARC.

McLay, G. (2005). *SEAWUN Benchmarking Survey for 2003 'Data Book'*. Retrieved from PSIREN website http://isslerhall.org/drupal/content/seawun-benchmarking-survey-2003-%E2%80%98data-book%E2%80%99.

McMahon, J. E. & Price, S. K. (2011). Water and energy interactions. *Annual Review of Environment and Resources, 36*, 163–191. doi:10.1146/annurev-environ-061110-103827.

Mekonnen, M. M. & Hoekstra, A. Y. (2011). The green, blue and grey water footprint of crops and derived crop products. *Hydrology and Earth System Sciences, 15*, 1577–1600. doi:10.5194/hess-15-1577-2011.

Ministry of Agriculture. (2014). Agricultural statistical yearbook 2014. Retrieved from Ministry of Agriculture website: http://www.moa.gov.sa.

Ministry of Water & Electricity. (2010). *Annual report 2010*. Retrieved from Ministry of Water & Electricity website: http://www.mowe.gov.sa.

Ministry of Water and Irrigation. (2015). *Energy efficiency and renewable energy policy for the Jordanian water sector*. Retrieved from Ministry of Water & Irrigation website http://www.mwi.gov.jo/sites/en-us/Hot%20Issues/Energy%20Efficiency%20and%20Renewable%20Energy%20Policy.pdf.

Moreno, M., Carrion, P. A., Planells, P., Ortega, J. F., & Tarjuelo, J. M. (2007). Measurement and improvement of the energy efficiency at pumping stations. *Biosystems Engineering, 98*, 479–486. doi:10.1016/j.biosystemseng.2007.09.005.

Multsch, S., Al-Rumaikhani, Y., Frede, H.-G., & Breuer, L. (2013). A site-specific agricultural water requirement and footprint estimator (SPARE:WATER 1.0). *Geoscientific Model Development, 6*, 1043–1059. doi:10.5194/gmd-6-1043-2013.

Natural Resources Canada Database. (2014). *Groundwater information [Data set]*. Retrieved February 10, 2014, from http://gin.gw-info.net/service/api_ngwds:gin/en/downloadmanager/dataset.html?package=waterwells.

Nelson, G. C., Robertson, R., Msangi, S., Zhu, T., Liao, X., & Jawajar, P. (2008). *Greenhouse gas mitigation: Issues for Indian agriculture*. Retrieved from IFPRI website http://www.ifpri.org/sites/default/files/publications/ifpridp00900.pdf.

Ouda, O. K. M. (2014). Water demand versus supply in Saudi Arabia: current and future challenges. *International Journal of Water Resources Development, 30*, 335–344. doi:10.1080/07900627.2013.837363.

Papapetrou, M., Wieghaus, M. & Biercamp, C. (2010). *Roadmap for the development of desalination powered by renewable energy (Deliverable 2.2)*. Retrieved from ProDes Project website: http://www.prodes-project.org/fileadmin/Files/ProDes_Road_map_on_line_version.pdf.

Pearce, G. K. (2007). UF/MF pre-treatment to RO in seawater and wastewater reuse applications: A comparison of energy costs. *Desalination, 222*, 66–73. doi:10.1016/j.desal.2007.05.029.

Peng, J. (2014). Market report: Developing desalination in China. Retrieved from http://www.waterworld.com/articles/wwi/print/volume-25/issue-6/regional-spotlight-asia-pacific/market-report-developing-desalination.html.

Pfeiffer, L. & Lin, C. Y. C. (2014). Does efficient irrigation technology lead to reduced groundwater extraction? Empirical evidence. *Journal of Environmental Economics and Management, 67*, 189–208. doi:10.1016/j.jeem.2013.12.002.

Plappally, A. K. & Lienhard, J. H. (2012). Energy requirements for water production, treatment, end use, reclamation and disposal. *Renewable and Sustainable Energy Reviews, 16*, 4818–4848. doi:10.1016/j.rser.2012.05.022.

Portela, L., & Cohim, E. (2011). *Avaliação da intensidade energética em sistemas de abastecimento de agua: O caso do SIAA de Feira de Santana* [Energy Intensity evaluation of water supply systems: The case of SIAA of Feira de Santana]. Retrieved from SEMIC website: http://www.xvisemic.esy.es/arquivos/sessao-iii/lais-portela.pdf.

Qatar General Electricity and Water Corporation. (2014). *Water statistics* [Data set]. Retrieved May 10, 2014, from www.km.com.qa.

Queensland Department of Natural Resources and Mines. (2014). *Groundwater database* [Data set]. Retrieved March 10, 2014, from http://www.dnrm.qld.gov.au/.

Raucher, R., Clements, J., Xu, P., Oxenford, J., Rutten, J. Choto, Z., & Reiss, R. (2010). *Guidelines for implementing seawater and brackish water desalination facilities*. Retrieved from Water Research Foundation website: http://www.waterrf.org/PublicReportLibrary/4078.pdf.

Reboucas, A.C. (1999). Groundwater resources in South America. Episodes: *Journal of International Geoscience, 22*, 232-237.

Renzoni, R. & Germanin, A. (2007). Life cycle assessment of water: From the pumping station to the wastewater treatment plant. *The International Journal of Life Cycle Assessment, 12*, 118–126.

Saatçı, M. A. (2015). *Içme Suyu Aritma Tesislerinde Proses Seçimi* [Drinking Water Treatment Plant Selection Process]. Retrieved from Suatiksu website: http://suatiksu.org/?wpfb_dl=59.

Sanders, K. & Webber, M. (2012). Evaluating the energy consumed for water use in the United States. *Environmental Research Letters, 7*, 034034. doi:10.1088/1748-9326/7/3/034034.

Shah, T. (2009). Climate change and groundwater: India's opportunities for mitigation and adaptation. *Environmental Research Letters, 4*, 035005. doi:10.1088/1748-9326/4/3/035005.

Shimizu, Y., Dejima, S., & Toyosada, K. (2012). The CO2 emission factor of water in Japan. *Water, 4*, 759–769. doi:10.3390/w4040759.

Sorrell, S. (2014). Energy substitution, technical change and rebound effects. *Energies, 7*, 2850–2873. doi:10.3390/en7052850.

Talozi, S., Al Sakaji, Y., & Altz-Stamm, A. (2015). Towards a water–energy–food nexus policy: realizing the blue and green virtual water of agriculture in Jordan. *International Journal of Water Resources Development, 31*, 461–482. doi:10.1080/07900627.2015.1040544.

Tao, X. (2012). *Energy consumption in wastewater treatment plants in China*. Unpublished manuscript retrieved from http://www.researchgate.net/profile/Tao_Xie11/publication/266146909_Energy_Consumption_in_Wastewater_Treatment_Plants_in_China/links/5428ce520cf238c6ea7cde91.pdf.

Tech Archival. (2014). *India water desalination: Assessment, opportunities & forecast up to 2018*. Retrieved from Market Reports on India website http://www.marketreportsonindia.com/energy-utility-market-research-reports-1082/india-water-desalination-assessment-opportunities-forecast-up-to-2018.html.

Tonner, J. (2008). *Barriers to thermal desalination in the United States (Desalination and Water Purification Research and Development Program Report No. 144)*. Retrieved from United States Bureau of Reclamation website: http://www.usbr.gov/research/AWT/reportpdfs/report144.pdf.

Trans Adriatic Pipeline-TAP. (2015). TAP's Environmental and social impact assessment (ESIA) process in Greece. Retrieved from http://www.tap-ag.com/our-commitment/to-the-environment/esia-greece.

United Nations Department of Economic and Social Affairs [UNDESA]. (2014). Water scarcity. Retrieved from http://www.un.org/waterforlifedecade/scarcity.shtml.

United Nations Statistics Division. (2015). *National Accounts Main Aggregates Database [Data set]*. Retrieved April 2, 2015, from http://unstats.un.org/unsd/snaama/dnllist.asp

Veera, G. G. (2011). Energy consumption and recovery in reverse osmosis. *Desalination and Water Treatment, 36*, 239–260. doi:10.5004/dwt.2011.2534.

Veera, G. G., Nirmalakhandan, N., & Deng, S. (2010). Renewable and sustainable approaches for desalination. *Renewable and Sustainable Energy Reviews, 14*, 2641–2654. doi:10.1016/j.rser.2010.06.008.

Venkatesh, G. (2011). *Systems performance analysis of Oslo's water and wastewater system (Unpublished Doctoral Thesis)*. Trondheim, Norway: Norwegian University of Science and Technology (NTNU).

Vieira, A. S., Beal, C. D., Ghisi, E., & Stewart, R. A. (2014). Energy intensity of rainwater harvesting systems: A review. *Renewable and Sustainable Energy Reviews, 34*, 225–242. doi:10.1016/j.rser.2014.03.012.

Vince, F. (2007). LCA tool for the environmental evaluation of potable water production. *Desalination, 220*, 37–56. doi:10.1016/j.desal.2007.01.021.

Wang, J., Routhausen, S. G. S. A., Conway, D., Zhang, L., Xiong, W., Holman, I. P., & Li, Y. (2012). China's water–energy nexus: greenhouse-gas emissions from groundwater use for agriculture. *Environmental Research Letters, 7*, 014035. doi:10.1088/1748-9326/7/1/014035.

Ward, F. A. & Pulido-Velazquez, M. (2008). Water conservation in irrigation can increase water use. *Proceedings of the National Academy of Sciences, 105*, 18215–18220. doi:10.1073/pnas.0805554105.

Water in the West. (2013). *Water and energy nexus: A Literature review*. Retrieved from Water in the West/Stanford Woods Institute for the Environment website: http://waterinthewest.stanford.edu/sites/default/files/Water-Energy_Lit_Review.pdf.

Water Resources Policy Division, Land and Water Bureau. (2006). *Outline of ground water management in Japan*. Retrieved from Network of Asian River Basin Organizations website: http://www.narbo.jp/narbo/event/materials/twwa03/tw03_09_03.pdf.

Wilkinson, R. C. (2000). *Methodology for analysis of the energy intensity of California's water systems, and an assessment of multiple potential benefits through integrated water-energy efficiency measures (Exploratory Research Project, Agreement No. 4910110)*. Retrieved from Stanford University website: http://large.stanford.edu/courses/2012/ph240/spearrin1/docs/wilkinson.pdf.

World Bank. (2015). Population, total [Data set]. Retrieved January 22, 2015, from http://data.worldbank.org/indicator/SP.POP.TOTL

World Health Organization & UNICEF. (2014). *Progress on drinking water and sanitation: 2014 update*. Retrieved from WHO/UNICEF Joint Monitoring Programme (JMP) for Water Supply and Sanitation website: http://www.wssinfo.org/fileadmin/user_upload/resources/JMP_report_2014_webEng.pdf.

World Wide Fund for Nature [WWF]. (2003). *The Real Cost of the Transfers: Analysis and socio-economic assessment of the Ebro transfers included in the Spanish National hydrological Plan, SNHP*. Retrieved from World Wide Fund for Nature website: http://assets.wwf.es/downloads/cost_transfers_ing_1.pdf.

Zilberman, D., Sproul, T., Rajagopal, D., Sexton, S., & Hellegers, P. (2008). Rising energy prices and economics of water in agriculture. *Water Policy, 1*, 11–21. doi:10.2166/wp.2008.049.

The energy trade-offs of adapting to a water-scarce future: case study of Los Angeles

Kelly Twomey Sanders

Sonny Astani Department of Civil and Environmental Engineering, University of Southern California, Los Angeles, USA

ABSTRACT

Increasing water competition, population growth and global climate change will intensify the tension between water and energy resources in arid climates of the world, since energy costs underscore the challenges facing water security in dry regions. In few places is the tension between water and energy resources more pronounced than in Los Angeles, California. This article analyzes the city's current water supply and estimates its future energy requirements based on water supply projections from the Los Angeles Department of Water and Power. Results suggest that while increasing local water management strategies could reduce the future energy intensity of the water supply, an increased reliance on water transfers could worsen its future energy intensity.

Introduction

Ensuring a safe and reliable water supply requires energy for pumping, treatment, distribution, end-use preparation and wastewater reclamation. In the absence of energy constraints, ubiquitous access to a safe and secure water supply would be achievable; water could be pumped any distance or treated regardless of initial water quality. However, water requires energy for treatment and pumping, and these energy demands are rising as clean, gravity-fed surface water sources become over-allocated, making integrated water and energy resource management increasingly important. Impaired sources of water, such as seawater, brackish water and other sources of contaminated water, require advanced treatment technologies to be suitable for drinking. Similarly, pumping water from deep groundwater aquifers and across long distances (i.e. interbasin transfers) require energy and advanced infrastructure that can increase overall costs and constrain development in poor regions (King, Stillwell, Twomey, & Webber, 2013; Sanders, King, Stillwell, & Webber, 2013; Sanders & Webber, 2012).

Nationally, the water system represents approximately 13% of total annual primary energy consumption in the US (Sanders & Webber, 2012). Although the majority of this energy use is for end-use preparation for activities such as water heating and industrial steam production, the treatment and delivery of water are still significant, representing about 4% of all water-related US energy consumption or about 1% of total US energy consumption

(Sanders & Webber, 2012). However, national averages do not adequately reflect the energy burden of many water-scarce regions, especially the burgeoning cities of the south-western US, which depend on energy-intensive sources such as interbasin transfers or desalinated ocean water to meet their demand. For example, California's water supply consists largely of interbasin transfers from water-rich (but population-sparse) regions to water-poor (but population-rich) regions. Consequently, statewide water-related energy usage in California represents approximately 19% of electricity consumption, of which about 25% is dedicated to energy pumping and treatment (CEC, 2006).

Los Angeles, the most populous city in California (3.9 million – Leslie, 2014) and the second-most populous city in the US, is also one of the nation's most water-stressed, making it an important case study in terms of evaluating the potential energy repercussions of future water management practices. The city's prosperity has been built on water sourced from distant watersheds to meet the demands of the city's population and economy; but securing this water supply, the majority of which originates hundreds of miles away, has come with large energy costs. In recent decades, Los Angeles has achieved steady population expansion without raising the cost of supplying water; however, future reductions in supply will challenge water managers to maintain an adequate water supply without undermining energy and environmental goals. Furthermore, rising pressures due to population and economic growth, greater competition over water supplies, prolonged drought, climate change, and heightened concerns over groundwater depletion and ecological uses of water have created the motivation to move towards new forms of water management practices (Barnett & Pierce, 2009; Cayan et al., 2010; Gleick, 2010).

This article examines Los Angeles's water management practices from an energy perspective. Business-as-usual water management practices are discussed, future trends are identified, and resulting energy impacts due to shifting water management strategies are quantified. The article also includes discussion of the social, economic and regulatory challenges facing the projected shifts.

An overview of water management in Los Angeles

The state of California has one of the most engineered water systems in the world. The majority of its water originates as snowpack in the northern third of the state, while the majority of the population lives in the southern third of the state, creating water management challenges. As a result, massive engineering projects have been developed to pump water from where it originates to population and agricultural centres. Los Angeles is served by three major aqueducts: the Colorado River Aqueduct, the California State Water Project (i.e. via the California Aqueduct), and the Los Angeles Aqueduct (LA Aqueduct). Each aqueduct varies in terms of the amount of water it delivers annually; however, collectively these sources can represent as much as 90% of Los Angeles' total water supply. Local groundwater generally makes up the majority of the remaining supply. Between 2009 and 2014, the water delivered by the Los Angeles Department of Water and Power (LA DWP) had the following sources in terms of percentage of total volume (LA DWP, 2015):

- LA Aqueduct (origin: Eastern Sierra Nevada), 37%
- State Water Project (origin: Bay Delta), 43%
- Colorado River Aqueduct (origin: Colorado River), 8%

- Local groundwater, 11%
- Recycled water, 1%

While these percentages reflect recent water supplies, the relative share of water from each of these sources varies considerably from year to year. During 2014, for example, water deliveries from the State Water Project wavered between zero and 5% of normal and were supplemented with larger-volume deliveries from the Colorado River Aqueduct and groundwater sources (Association of California Water Agencies, 2014; Boxal, 2014). The following sections detail these water supply sources.

Surface water deliveries via aqueducts

Of the three major aqueducts serving the city of Los Angeles, the LA Aqueduct is the only one managed by LA DWP. Water imported via the State Water Project and the Colorado River Aqueduct is managed by the Metropolitan Water District (MWD) of Southern California, which then distributes water to LA DWP and its other member agencies across Southern California.

The LA Aqueduct, the first aqueduct built to convey water to Los Angeles, was completed in 1913. Although the pipeline spans hundreds of miles, it is gravity-fed from source to treatment facility, eliminating any energy demand until treatment. In fact, the aqueduct is actually a net energy producer, as it generates hydroelectric power along its course. A second LA Aqueduct was added decades later, in 1970, so that the city could divert its full allocation of water from Mono Basin, the Owens River, and the Owens Valley (LA DWP, 2010). The LA Aqueduct has been contentious from the beginning due to concerns regarding the social, economic and environmental repercussions of depletion of water at the aqueduct's source (Stroshane, 2008). Consequently, water deliveries from the LA Aqueduct have been substantially curtailed, to less than half of historical annual diversion volumes, in recent years. A large fraction of the water (approximately 47%) that would otherwise be diverted to Los Angeles is now used for dust control and other environmental mitigation and enhancement projects to ameliorate the ecosystem impacts created by the diversion projects (Hughes, Pincetl, & Boone, 2013; LA DWP, 2010). Approximately 39% of the remainder of the water allocated to Los Angeles from the LA Aqueduct reaches the city, while the remaining 14% is lost to evaporation and infiltration (LA DWP, 2010).

MWD manages water imports from the Colorado River Aqueduct and the State Water Project, which originate in the Colorado River and Northern California, respectively. The 242-mile Colorado River Aqueduct first delivered water in 1941. It begins on the Colorado River near Parker, Arizona, and ends at the Cajalco Reservoir, near Riverside, California (Bissell, 1939). While MWD owns the Colorado River Aqueduct, the State Water Project is owned by the California Department of Water Resources, and allocates water to MWD based on water availability. The entire State Water Project has 701 miles of canals and pipelines and is the single largest user of electricity in the state (Lofman, Petersen, & Bower, 2002). Water is gravity-fed until it reaches the Banks Pumping Plant on the California Aqueduct, where it is pumped approximately 440 miles to Southern California's distribution networks. The volumes of water subject to purchase by MWD from the State Water Project were reduced every year between 2007 and 2014. Much of these curtailments is a result of the enforcement of the Endangered Species Act because of concerns over threatened and endangered species, such as the delta smelt (LA DWP, 2010; Mo, Wang, & Zimmerman, 2014). More recently, drought has constrained the water available for delivery to Southern California.

Groundwater pumping

Over the past decade, groundwater has represented about 12% of the city's water supply, but in dry years it can represent as much as 30% (LA DWP, 2010). The city has water rights to five groundwater basins – the San Fernando, Sylmar, Eagle Rock, Central, and West Coast Basins – however, groundwater contamination has reduced the city's ability to pump local groundwater in recent years (LA DWP, 2010). Currently there are nine groundwater management agencies, which manage 24 adjudicated groundwater basins in Southern California, whose boundaries are defined by court. These agencies are regulated in terms of how much water they can extract from the aquifer, in efforts to avoid over-extraction and depletion, seawater intrusion, and decreasing local water supply (CWF, 2014).

Groundwater recharge, that is the pumping of water into groundwater aquifers for temporary storage, is utilized in Southern California for storage and to protect costal aquifers from seawater intrusion. Potable water is typically mixed with recycled water treated to a quality that satisfies indirect potable reuse standards, before being pumped into a receiving aquifer (Lazarova, Oelker, & Won, 2012).

Water recycling

Water recycling currently represents a small (but growing) fraction of Los Angeles's water supply. There are a number of recycled-water projects, including the Terminal Island Water Reclamation Plant, Los Angeles/Glendale Water Reclamation Plant, Donald C. Tillman Water Reclamation Plant, and Edward C. Little Water Recycling Facility, which supply the city with recycled water (LA DWP, 2010). The Little plant, in El Segundo, Los Angeles County, is considered one of the most innovative water treatment facilities in the US and has plans to expand. Operated by West Basin Municipal Water District, the plant receives secondary treated wastewater effluent from the city of Los Angeles's Hyperion Wastewater Treatment Facility and treats it to a quality consistent with its intended end use (Lazarova et al., 2012). In total, the facility and its satellite locations treat the effluent to produce five qualities of treated recycled water, an average of 40 million gallons of recycled water per day. Although the effluent discharged from Hyperion is suitable for discharge into the ocean, biological, organic and inorganic materials still remain in the wastewater effluent that must be removed for it to be suitable for other reuse applications under the current regulatory regime (Brant, Kwan, Daniel, & Valencia, 2010). By tailoring recycled water quality to end use (primarily irrigation, industrial uses and groundwater recharge), the facility has flexibility in its operational processes and the environmental and economic costs of water recycling can be reduced (Hering, Waite, Luthy, Drewes, & Sedlak, 2013).

Despite the potential benefits water recycling can offer a community, especially in water-scarce regions such as Southern California, there are trade-offs that thwart facility expansion. In the absence of direct potable reuse (i.e. recycled water used for drinking), securing a customer base to receive nonpotable streams of water can be difficult; customers with appropriate nonpotable water needs must be identified and new piping infrastructure must be built. 'Purple' pipe, a third delivery piping infrastructure, is generally used to deliver recycled water that is of lower quality than potable water but higher quality than wastewater (Stillwell et al., 2011). In Southern California, the majority of recycled water goes to large industrial or power facilities that require enough water to justify the cost of building or extending a new pipeline (Sanders, 2015). Connecting a single residential or commercial customer location is often cost-prohibitive in the absence of regulation or governmental assistance. Although recycled-water infrastructure is expanding in Los Angeles, it can take years for purple-pipe infrastructure to reach the customers that would like access to the resource. In addition to the building of new delivery systems, the process of water recycling itself is also relatively expensive compared to other options, most notably due to energy consumption and membrane fouling (Brant et al., 2010). Furthermore, water recycling facilities are typically at relatively low elevations in a distribution network, so recycled water often has to be pumped uphill to reach end users (Stillwell et al., 2011).

Water conservation

Today, Angelenos use approximately 0.450 m^3 per person per day, down from 0.650 m^3 per person per day in the 1980s (Hughes et al., 2013). Much of this water reduction has been a function of an aggressive indoor water conservation campaign to install low-flow toilets, shower heads and faucets, high-efficiency appliances, and closed-loop (i.e. recirculating) cooling towers, accompanied by changes to the city's water code to promote low-water fixtures in new build-outs (LA DWP, 2010). Tiered water pricing structures were also implemented to reduce water waste (City of Los Angeles, 2008; Hughes et al., 2013). While the city has been successful in markedly reducing per capita water usage, outdoor water usage remains high. Outdoor water usage represents approximately 54% of LA DWP's single-family-residence water deliveries; for multi-family construction, it represents about 32%. Across its entire services area, between 2004 and 2007, about 39% of LA DWP's supplied water went for outdoor irrigation (LA DWP, 2010).

Energy implications

In most regions, clean surface water supplies are utilized before more costly options. This trend implies that the next marginal water source is generally more expensive (and more energy-intensive) than baseline water supplies (Sanders & Webber, 2012). However, since most of California's population growth has been predicated on its ability to move water from population-sparse and water-rich regions to population-dense and water-poor regions, shifting to more local water sources in Southern California might incur energy benefits compared to business-as-usual water management, depending on the technologies utilized as well as future water demand. Table 1 provides energy-intensity estimates from the literature for delivering, treating and reclaiming water in Southern California.

According to its 2010 Urban Water Management Plan, LA DWP expects annual water demand to increase by 18–26% between 2010 and 2035, depending on the level of conservation strategies implemented (LA DWP, 2010). However, the water management solutions of the future are likely to look different from the water systems of the past. The economic practicality of building new large surface water storage projects in the state is believed by most to be very low (Gleick, 2008). More distributed water storage solutions, such as rainwater collection and stormwater management, offer a great deal of potential in terms of building a local, sustainable supply (Gleick, 2008). Consequently, Los Angeles is moving towards goals that increase recycled water and water conservation, expand stormwater capture, improve

Table 1. Energy intensities of selected water supplies in California

California: Specific Energy Intensity Estimates from the Literature	Low MJ per m^3	High MJ per m^3	Sources
Raw water supply pumping			
State Water Project – West Branch	7.52		LA DWP, 2010
State Water Project – East Branch	9.44		LA DWP, 2010
Colorado River Aqueduct	5.48		LA DWP, 2010
Los Angeles Aqueduct*	0		LA DWP, 2010
Local groundwater supplies	0.5	2.48	Plappally & Lienhard V, 2012; R. Wilkinson, 2000
Treatment and distribution			
Potable treatment and distribution	0.07	0.25	Dale, 2004; Plappally & Lienhard V, 2012; R. C. Wilkinson, 2007; R. Wilkinson, 2000
Brackish water RO	1.18	9.25	Plappally & Lienhard V, 2012
Seawater desalination	9.29	19.76	Plappally & Lienhard V, 2012
Wastewater treatment and collection			
Wastewater collection	1.05	4.91	Plappally & Lienhard V, 2012; R. Wilkinson, 2000
Recycled water treatment and distribution			
Title 22 quality	1.14	3.25	Plappally & Lienhard V, 2012; R. C. Wilkinson, 2007
Intermediate quality	2.63	5.17	Plappally & Lienhard V, 2012
High quality	7.84	12.37	Dale, 2004; Plappally & Lienhard V, 2012; WBMWD, 2014; R. C. Wilkinson, 2007
Recycled water treatment and distribution			
Groundwater recharge	1.02	1.8	R. C. Wilkinson, 2007

Notes: *The LA Aqueduct is a net energy producer because it is gravity-fed and generates hydroelectric power at several locations along its course (LA DWP, 2010).

groundwater management, and enhance green building initiatives, while better protecting ecosystems (California Natural Resources Agency, 2015; Cousins & Newell, 2015).

Methodology

The energy required to supply water to the city of Los Angeles varies considerably from year to year, and will continue to shift as new water management strategies are explored. To anticipate the energy consumption trends of Los Angeles' future water management solutions, the energy demands of water supply projection defined in the LA DWP's Urban Water Management Plan were calculated (LA DWP, 2010). Table 2 details LA DWP water supplies in the FY2009-10 baseline year, while Table 3 details three LA DWP water demand and supply projections for FY2034-35 based on varying weather scenarios (an average weather year, a single critical dry year, and five cumulative dry years). The three FY2034-35 projections are nearly identical except for total anticipated annual volumetric water demand, and the

Table 2. Historical water supply volumes and water supply energy intensity estimates for the LA DWP service territory during FY2009-2010.

			Water allocation per category in 2009-2010	Raw water supply pumping		Potable water treatment and distribution		Wastewater treatment, collection and discharge		Recycled water treatment and distribution	
				Low ($EI_{1,L}$)	High ($EI_{1,H}$)	Low ($EI_{2,L}$)	High ($EI_{2,H}$)	Low ($EI_{3,L}$)	High ($EI_{3,H}$)	Low ($EI_{4,L}$)	High ($EI_{4,H}$)
			Historical volume, V_m	$F_{1,L}=1$	$F_{1,H}=1$	$F_{2,L}=1$	$F_{2,H}=1$	$F_{3,L}=0.3$	$F_{3,H}=0.8$	$F_{4,L}=1$	$F_{4,H}=1$
			m^3	MJ per m^3							
LA DWP potable water supplies	Metropolitan Water District imports (77% SWP West, 13% SWP East, 10% Colorado River)	V_1	3.25E+08	7.60		0.07	0.25	1.05	4.91		
	LA Aqueduct	V_2	2.46E+08	0		0.07	0.25	1.05	4.91		
	Additional future water transfers	V_3	0.00E+00								
LA DWP alternative water supplies	Local surface supplies	V_4	0.00E+00								
	Local groundwater	V_5	9.50E+07	0.50	2.48	0.07	0.25	1.05	4.91		
	Recycled water: irrigation and industrial	V_6	8.27E+06							1.14	12.37
	Recycled water: groundwater replenishment	V_7	0.00E+00								
	Stormwater capture and reuse	V_8	0.00E+00								
	Stormwater for groundwater recharge	V_9	0.00E+00								
	Conservation	V_{10}	1.01E+07	n/a	n/a	n/a	n/a	n/a	n/a	n/a	n/a
	Total (less conservation)	V_{DWP}	6.75E+08	—	—	—	—	—	—	—	—
	Total		6.85E+08	—	—	—	—	—	—	—	—

Source: Volume data are based on LA DWP (2010, Table 11F) historical and projection data. Energy Intensity data are from the literature cited in Table 1.

Table 3. LA DWP water supply projections for the year FY2034-2035 and anticipated energy intensities.

Los Angeles DWP water demand and supply projections for FY 2034-35	V_m	Water allocation per category			Raw water supply		Potable water treatment and distribution		Wastewater treatment, collection and discharge		Recycled water treatment and distribution	
		Average weather year	Single critical dry year[a]	Multiple dry years[b]	Low ($EI_{1,L}$) $F_{1,L}=1$	High ($EI_{1,H}$) $F_{1,H}=1$	Low ($EI_{2,L}$) $F_{2,L}=1$	High ($EI_{2,H}$) $F_{2,H}=1$	Low ($EI_{3,L}$) $F_{3,L}=0.3$	High ($EI_{3,H}$) $F_{3,H}=0.8$	Low ($EI_{4,L}$) $F_{4,L}=1$	High ($EI_{4,H}$) $F_{4,H}=1$
		m^3			MJ per m^3							
Metropolitan Water District imports[c] (Low: 100% Colorado River; High: 100% SWP East)	V_1	2.07E+08	5.03E+08	3.95E+08	5.84	9.44	0.07	0.25	1.05	4.91		
LA Aqueduct[d]	V_2	3.01E+08	5.79E+07	1.30E+08	0		0.07	0.25	1.05	4.91		
Additional future water transfers[e]	V_3	4.93E+07	4.93E+07	4.93E+07	0.50	9.44	0.07	0.25	1.05	4.91		
Local surface supplies	V_4	0.00E+00	0.00E+00	0.00E+00	0.12	0.13	0.07	0.25	1.05	4.91		
Local groundwater[f]	V_5	1.36E+08	1.36E+08	1.36E+08	0.50	2.48	0.07	0.25	1.05	4.91		
Recycled water: irrigation and industrial[g]	V_6	3.58E+07	3.58E+07	3.58E+07							1.14	12.37
Recycled water: groundwater replenishment[g]	V_7	3.70E+07	3.70E+07	3.70E+07							1.02	1.80
Stormwater capture and reuse[h]	V_8	1.23E+07	1.23E+07	1.23E+07							0	12.37
Stormwater for groundwater recharge[h]	V_9	1.85E+07	1.85E+07	1.85E+07							1.02	1.80
Conservation[i]	V_{10}	7.94E+07	7.94E+07	7.94E+07	n/a	n/a	n/a	n/a	n/a	n/a	n/a	n/a
Total (less conservation)	V_{DWP}	7.97E+08	8.50E+08	8.15E+08	—	—	—	—	—	—	—	—
Total		8.77E+08	9.29E+08	8.94E+08	—	—	—	—	—	—	—	—

Notes: [a]Defined as a repeat of the 1990–1991 drought. [b]Defined as a five-year dry period repeating the conditions experienced during FY1988-89 through FY1992-93. [c]MWD anticipates 100% supply reliability through 2035 for all members due to increased storage and new supplies; DWP assumes imports are available as needed. [d]Assumed by LA DWP to decline 0.17%/y due to climate change and to fluctuate substantially in dry years. [e]Anticipated new supplies including an interconnection between the State Water Project water and the Los Angeles Aqueduct; impacts to energy intensity are unclear. [f]Assumed constant regardless of scenario as a conservative estimate to avoid groundwater aquifer depletion. [g]Assumed constant since available supplies are moderated by system capacity, but are not affected in drought years. [h]Considered a "potential supply option" for the future, as stormwater reuse and infiltration is not currently utilized; otherwise, more MWD imports will be needed. [i]No details regarding conservation strategies are given, so energy intensity is not assessed in this analysis.

Source: Volume data are based on LA DWP (2010) historical and projection data. Energy Intensity data are from the literature cited in Table 1.

imported water volumes received from MWD and the LA Aqueduct to meet this demand (LA DWP, 2010).

The total annual energy consumed for moving, treating and reclaiming the city of Los Angeles' water supply is estimated for FY2009-10 and the three FY2034-35 projections using Equations (1) and (2), representing the lower and upper bound of annual energy consumption, respectively. The average energy intensity (in terms of electricity consumption per unit volume) of the entire LA DWP water supply (V_{DWP}) is estimated with Equation (3). In Equations (1) and (2), V_m represents the historical or projected water volume supplied by water supply source, m, based on LA DWP data and projections. EI_n values reflect the lower (subscript L) or higher (subscript H) energy-intensity estimates for each lifecycle stage n. F_n represents the fraction of each water supply volume, V_m, subject to pumping or treatment at lifecycle stage n. All values of V_m, V_{DWP}, EI_n and F_n are defined in Tables 2 and 3 for each respective scenario.

$$E_{DWP,L} = \sum_{m}^{10} \sum_{n}^{4} (V_m \times (F_{n,L} \times EI_{n,L})) \tag{1}$$

$$E_{DWP,H} = \sum_{m}^{10} \sum_{n}^{4} (V_m \times (F_{n,H} \times EI_{n,H})) \tag{2}$$

$$EI_{DWP,H\ or\ L} = \frac{E_{DWP,H\ or\ L}}{V_{DWP}} \tag{3}$$

Since Los Angeles utilizes a number of imported water supplies, water treatment facilities, wastewater treatment and recycling facilities, which vary considerably from year to year in terms of water volume supplied or treated, the bounded analysis offers insight into the potential range of energy that might be required based on water supply and treatment decisions. In FY2009-10, MWD imports were 10% and 90% from the Colorado River Aqueduct and the State Water Project, respectively (LA DWP, 2010). However, MWD water supplies vary greatly from year to year. For example, water imports to Los Angeles from the State Water Project were reduced to 5% in 2014, reflecting uncertainty in MWD's imported water supply mix moving forward (Association of California Water Agencies, 2014). Thus, in the FY2034-35 projection scenarios, the Low case reflects imports from MWD being 100% from the Colorado River Aqueduct, while the High case assumes 100% from the State Water Project, in order to bound energy consumption projections for a low and high estimate, respectively. Although it is likely that imports will represent a range of these sources, these assumptions make it possible to analyze the possible range of energy consumption for importing these supplies.

Several other assumptions were made to calculate energy consumption estimates. It is assumed that 100% of each potable or alternative water supply option reaches the first stage of its anticipated treatment (i.e., the volume of water represents volume delivered). However, in the case of the potable water supply, the fraction of water that flows through the sewer system to a wastewater facility depends on losses along the way. In public water supply territories that have high leakage rates and/or use a large fraction of potable water for outdoor irrigation, the fraction of water flowing to a wastewater facility will be smaller than areas that have smaller fractional losses (Sanders & Webber, 2012). This consideration is reflected in the bounded analysis, which assumes that 30% and 80% of the original potable water supply reaches a wastewater treatment facility in the lower- and upper-bound scenarios, respectively.

Estimating the energy consumed for water also requires assumptions regarding the quality to which each source will be treated and how the water will be pumped through the distribution network. For example, potable water systems typically use gravity to move water from higher to lower elevations in the distribution system. Recycled water, by contrast, is typically treated at facilities at low elevations in the distribution network and then must be pumped against gravity to its end user (Sanders & Webber, 2012; Stillwell et al., 2011). For alternative water supplies, there is a great deal of uncertainty in regard to future treatment standards, pumping requirements and end-use quality. For example, collecting and utilizing stormwater for water reuse or groundwater recharge is not currently practiced by LA DWP, so it is unclear whether these supplies will be subject to high potable water standards or be used on-site for nonpotable uses requiring little to no energy. Similarly, the energy requirements of reclaimed water are subject to large changes over time, especially if direct potable reuse is implemented.

Recycled water sources for irrigation and industrial uses are assumed to be treated to tertiary-quality (Title 22) and high-quality (i.e. double-pass reverse osmosis) standards for the Low and High cases, respectively. (Although it is more likely that these recycled supplies will trend towards lesser treatment, some industrial applications require water of quality higher than potable standards; Sanders & Webber, 2012.) Water collected for stormwater management was assumed to have no energy input for the Low scenario and to be treated with double-pass reverse osmosis in the High scenario. Stormwater systems are often passive and require no energy input, but a high bound is assumed here in case these supplies are used to supplement either direct or indirect potable reuse in the future.

Results and discussion

Table 4 summarizes the results of the bounded energy analysis of the LA DWP water supply for the FY2009-10 baseline year and three FY2034-35 projection scenarios. The FY2034-35 water demand and supply projections vary depending on whether the system is subject to an average, critical-dry-year or prolonged-drought scenario. The LA DWP owns and operates the LA Aqueduct and therefore has depended heavily on it in the past. However, water from this source varies substantially from year to year and has decreased in recent years because of environmental regulations reducing the amount of water LA DWP can pump to the city from its source. The LA DWP's projections reflect large uncertainties in the water available from the LA Aqueduct over time. The water supplied by the LA Aqueduct during FY2034-35 in an average-weather year, for example, is anticipated to be 80% more than in a critical-dry year and 57% more than for a scenario with multiple dry years preceding it. Even in the average-weather scenario, climate change will reduce the water available from this source compared to current availability (LA DWP, 2010). Since the LA DWP projections for all other water supplies, besides MWD imports, are assumed to be the same across these scenarios, any reduction in the LA Aqueduct water must be met by increases in MWD imports or reductions in the total water supplied. Thus, when the water supplied by the LA Aqueduct, which requires no energy for pumping, is significantly reduced, it is replaced with water that is pumped hundreds of miles from Northern California or the Colorado River, significantly

Table 4 Lower and upper bound energy consumption and energy intensity estimates for the LA DWP FY 2009-2010 historical case and three FY2034-2035 projection scenarios

	FY 2009-2010 historical year		FY 2034-2035 average-weather year		FY 2034-2035 single dry-weather year		FY 2034-2035 multiple dry years preceeding 2034	
Energy consumption for water supplies (less conservation)	Low ($E_{DWP,L}$)	High ($E_{DWP,H}$)	Low ($E_{DWP,L}$)	High ($E_{DWP,H}$)	Low ($E_{DWP,L}$)	High ($E_{DWP,H}$)	Low ($E_{DWP,L}$)	High ($E_{DWP,H}$)
	MJ		MJ		MJ		MJ	
Metropolitan Water District imports	2.60E+09	3.84E+09	1.29E+09	2.82E+09	3.13E+09	6.85E+09	2.46E+09	5.78E+09
LA Aqueduct	9.52E+07	1.03E+09	1.16E+08	1.26E+09	2.24E+07	2.42E+08	5.04E+07	5.46E+08
Additional future water transfers	0.00E+00	0.00E+00	4.39E+07	6.72E+08	4.39E+07	6.72E+08	4.39E+07	6.72E+08
Local surface supplies	0.00E+00	0.00E+00	0.00E+00	0.00E+00	0.00E+00	0.00E+00	0.00E+00	0.00E+00
Local groundwater	8.45E+07	6.33E+08	1.21E+08	9.08E+08	1.21E+08	9.08E+08	1.21E+08	9.08E+08
Recycled water: irrigation and industrial	9.44E+06	1.02E+08	4.08E+07	4.43E+08	4.08E+07	4.43E+08	4.08E+07	4.43E+08
Recycled water: groundwater replenishment	0.00E+00	0.00E+00	3.78E+07	6.66E+07	3.78E+07	6.66E+07	3.78E+07	6.66E+07
Stormwater capture and reuse	0.00E+00	0.00E+00	0.00E+00	1.53E+08	0.00E+00	1.53E+08	0.00E+00	1.53E+08
Stormwater for groundwater recharge	0.00E+00	0.00E+00	1.89E+07	3.33E+07	1.89E+07	3.33E+07	1.89E+07	3.33E+07
Conservation	n/a	n/a	n/a	n/a	n/a	n/a	n/a	n/a
Total energy consumption for water supplies (less conservation)	2.79E+09	5.60E+09	1.67E+09	6.36E+09	3.41E+09	9.37E+09	2.77E+09	8.60E+09
Averagy energy intensity of annual water supply (less conservation)	Low ($EI_{DWP,L}$)	High ($EI_{DWP,H}$)	Low ($EI_{DWP,L}$)	High ($EI_{DWP,H}$)	Low ($EI_{DWP,L}$)	High ($EI_{DWP,H}$)	Low ($EI_{DWP,L}$)	High ($EI_{DWP,H}$)
	MJ per m^3		MJ per m^3		MJ per m^3		MJ per m^3	
	4.1	8.3	1.9	7.3	3.7	10.1	3.1	9.6

increasing the energy intensity of the water supply in the absence of large cuts to water demand. This trend is reflected in results (Table 4) for the projected energy intensity of water supplies in FY2034-35 for the two dry-year scenarios, compared to the average-weather-year scenario.

Results in Table 4 indicate that the energy intensity of water supplies could also decrease due to the increased usage of alternative water supplies. However, the ultimate impacts to energy consumption by exploiting these supplies varies considerably depending on end-use quality targets. Furthermore, stormwater capture supplies are considered 'potential supplies', so if these supplies are not exploited by LA DWP in the future as indicated in Table 3, energy-intensive imports from MWD will be required to replace them. Although recycled water can be more energy-intensive than imported water when subject to the most energy-intensive water treatments, there is value in increasing local water supplies, especially in a regime in which distant water sources derived from Northern California and Rocky Mountain snowpack are in decline. Thus, energy intensity is not the only valuable metric when considering the future of Los Angeles' water supplies – resilience to disruptions must also be considered. Matching water quality with water use offers a valuable opportunity to achieve both targets: reducing energy intensity while increasing local supply. However, large, expensive infrastructure projects would be required to extend nonpotable water delivery, making these changes difficult. That being said, Los Angeles' water system is aging and will require extensive repair in the near term. Nonpotable distribution systems should be installed as existing potable and sewer pipes are replaced.

Notably omitted from the future water projections included in Table 3 is desalination. Since the majority of urban water demand in California occurs in coastal cities, seawater desalination is expected to increase. However, LA DWP has not included desalination as a part of its expected water supply portfolio through 2035 due to the economic and environmental challenges of siting desalination facilities in comparison to other supply options such as water recycling.

A changing paradigm for Los Angeles water management

There are various reasons that motivate changes to Southern California's business-as-usual water management. By 2025, the city of Los Angeles expects over 360,000 new residents to move to the city, an increase of 10% over its 2013 population (Cousins & Newell, 2015). Aside from Los Angeles's growth, population growth in other watersheds fed by Sierra Nevada snowpack and the Colorado River is also anticipated, adding significant pressure to these declining water sources.

Severely compounding issues of population growth in water-scarce regions of the US is the fact that existing sources of water are already in decline. And climate change models suggest that the snowpack regions supplying the city's imported water will continue to decline. Snowpack in the Sierras (the major water source for the State Water Project and LA Aqueduct) might shrink 25 – 40% by 2050, according to climate models (Cousins & Newell, 2015). The Colorado River (the water source for the Colorado River Aqueduct) is already approaching complete allocation, and climate models suggest that runoff will fall 10–30% by 2050 (Barnett & Pierce, 2009).

While LA DWP projections assume a 0.17% decline per year in water available for transport via the LA Aqueduct due to climate change, its projections regarding future drought are based on prior drought events. The one-year critical drought scenario is based on the California drought spanning 1990–91, while the multi-year drought scenario is based on the five-year dry period from 1988 to 1993 (LA DWP, 2010). The current California Drought, beginning in 2012 and continuing through 2015 (and possibly far into the future), is considered the most severe in 1200 years, with accumulated moisture deficits worse than any previous period (Griffin & Anchukaitis, 2014). Statewide snow water equivalents were 21% of normal, as of 6 February 2015 (CA DWR, 2015a). The severity of this drought has been driven, in part, by record high temperatures, which are expected to continue to rise with climate change (Griffin & Anchukaitis, 2014; Karl, Melillo, & Peterson, 2009). Therefore, recent drought events have already been far worse than the dry period of 1988–93, suggesting that planning efforts based on historical conditions might not be adequate for anticipating future extreme events.

Aside from physical water availability, existing imported-water sources serving Los Angeles have already been curtailed in recent years because of regulatory decisions to protect the environment in the case of the LA Aqueduct, water shortage and environmental concerns in the case of the State Water Project, and equitable distribution of water in the case of the Colorado River (LA DWP, 2010; Pulido-Velazquez, Jenkins, & Lund, 2004). As surface water allocations have shrunk, groundwater depletion has increased, exacerbating land subsidence and salinity encroachment. Despite the creation of nine groundwater agencies (mostly in Southern California), the majority of California's groundwater resources are not monitored. New methods of satellite observation have made it possible to detect severe groundwater depletion in much of California (Famiglietti et al., 2011; Konikow, 2015). However, with the passage of the 2014 Sustainable Groundwater Management Act, the state is establishing a new and improved structure for managing the state's groundwater resources (CA DWR, 2015b).

Technical, economic and regulatory challenges

Sustainable water management solutions are not a function of technological shifts and innovation alone. Regulatory, economic and social issues will also impact the future of Los Angeles's water management strategies. The ways these nontechnical factors will impact the water supply and its energy intensity are also hard to predict. As technology improves, and as our understanding of the public health consequences of emerging contaminants improves, regulatory standards generally also improve. This trend suggests that the drinking-water supply sources of the future are likely to require treatment to a higher end-use quality than those of the present.

Changes in regulations could also warrant energy improvements. Currently, the US generally requires that all water supplied through the public water supply meet potable water standards (Sanders & Webber, 2012). In reality, only a small fraction of the urban water supply warrants water of such high quality. Water used to irrigate, for example, could be of lower quality without compromising its utility to its end user. Recycled water offers an opportunity to optimally match water quality with water use. However, to do this effectively, existing regulations and codes will have to be modified to allow the plumbing modifications necessary for on-site water recycling, for example. Existing codes in most regions of the United States prohibit direct potable reuse, i.e. reconditioning treated wastewater effluent to a quality acceptable for drinking (King et al., 2013; Stillwell et al., 2011). For example, the California

Health and Safety code (§116551, 1998) states that the department "shall not issue a permit to a public water system or amend a valid existing permit for the use of a reservoir as a source of supply that is directly augmented with recycled water". However, in 2013 the State of California's Water Code (§§13560-13569, 2013) was amended, stating that "on or before December 31, 2016, the department, in consultation with the state board, shall investigate and report to the Legislature on the feasibility of developing uniform water recycling criteria for direct potable reuse", marking increased interest in this resource. Without direct potable water reuse, recycled water cannot be mixed directly into the potable drinking water supply without an environmental buffer, as in the case of indirect potable water reuse, e.g. groundwater replenishment (Wetterau, Chalmers, Liu, & Pearce, 2013). Current regulations that restrict the practice of direct potable reuse greatly restrict the utilization of local water sources such as recycled water and captured stormwater on the large scale, since nonpotable water requires an additional purple-pipe distribution system that is expensive to build (Hering et al., 2013).

Increasing local water management solutions can also provide environmental benefits. Stormwater capture, for example, is expected to increase substantially by 2035 through distributed and centralized projects, which will simultaneously reduce the amount of water flowing to the ocean and increase the recharge of local aquifers (LA DWP, 2010). Aside from a few small projects, effective stormwater management is at present nearly nonexistent. The vast majority of stormwater runoff is diverted to storm drains, where it is then channelled to the ocean. Not only is this local water supply wasted as a potential supplement to the city's current water budget, but as stormwater flows towards the ocean it acquires contaminants such as grease, oils, bacteria, trash and heavy metals that ultimately pollute the ocean. Furthermore, with increased urbanization, the natural recharge of aquifers has also been thwarted, since more of the city's surface is covered with impermeable surfaces. In addition to centralized and decentralized stormwater management, rain barrels, cisterns, rain gardens and neighbourhood recharge programs will continue to grow as a part of Los Angeles's urban water management strategy (LA DWP, 2010).

In addition to supply-side water management, there are large improvements that can address demand-side solutions. Market-based solutions in the water sector are currently very weak compared to other sectors of the economy, since water is viewed as a resource that should be universally accessible to all. Though it is politically unfavourable, water must be properly priced to provoke meaningful changes in behaviour. Without a proper price, customers are not incentivized to reduce usage, and infrastructure cannot be properly maintained (Gleick, 2008; Hung & Chie, 2012; Mehan & Kline, 2012). Although equitable access to water must be ensured, more aggressive tiered water pricing structures, for example, could ensure that water is accessible to equitably meet society's basic needs but would also reduce the amount of water used for wasteful purposes. A proper price on water would also reduce waste within distribution systems. Most of the water system in Southern California is over 100 years old, so leaks and pipe breaks contribute to system losses that would be better managed if utilities had budgets commensurate with maintaining ageing infrastructure.

There are other market-based solutions. Although water markets in California exist, more functional water markets can facilitate more efficient transfers of water, especially from agricultural to urban entities, and especially in combination with improved conjunctive uses of water (Gleick, 2008; Pulido-Velazquez et al., 2004). The California Water Action Plan recognizes the need for more effective conjunctive management of surface and groundwater supplies to restore contaminated groundwater supplies and increase storage for dry years (California Natural Resources Agency, 2015). MWD has negotiated agreements to increase water deliveries in wet years to recharge groundwater aquifers so that storage is available in dry years when deliveries might be curtailed, as they were in 2014 (Pulido-Velazquez et al., 2004). This conjunctive use of water adds operational flexibility so that water transfers are put towards their highest economic use (California Natural Resources Agency, 2015). More generally, incorporating more holistic water planning through integrated water resource management strategies that recognize the interdependency of water and all sectors of the economy, while balancing multi-criteria objectives (environmental, climate change, economics, etc.), will be increasingly important moving forward (Cooley & Gleick, 2011; King et al., 2013).

Conclusion

Los Angeles has made progress in sustainable water management efforts in recent years. However, population and economic growth, climate change and drought, and increasing water competition will continue to challenge the city. This article reviews recent water supply projections made by LA DWP and anticipates the potential energy requirements of those estimates. Results suggest that increasing local water management strategies such as recycled water and stormwater management can offer benefits over business-as-usual water-related energy consumption, specifically by reducing water imports. However, if imports remain high and/or local water supplies are treated with highly energy-intensive treatment processes, such as reverse osmosis, water-related energy consumption could go up, especially in the absence of significant efforts to reduce water consumption.

Disclosure statement

No potential conflict of interest was reported by the author.

References

California Natural Resources Agency. (2015). *California water action plan implementation report 2014–2018*. Report. Retrieved from: http://resources.ca.gov/california_water_action_plan/implementation_report/

Association of California Water Agencies. (2014, December 1). Initial state water project allocation for 2015 Set at 10 Percent. *Water News*. Sacramento, CA. Retrieved from http://www.acwa.com/news/water-supply-challenges/initial-state-water-project-allocation-2015-set-10-percent

Barnett, T. P., & Pierce, D. W. (2009). Sustainable water deliveries from the Colorado River in a changing climate. *Proceedings of the National Academy of Sciences, 106*, 7334–7338. doi:10.1073/pnas.0812762106.

Bissell, C. A. (Ed.). (1939). *The Metropolitan water district of Southern California: History and first annual report* (1st ed.). Los Angeles, CA: Haynes Corporation. Retrieved from http://www.mwdh2o.com/mwdh2o/pages/about/AR/AR1928.html

Boxal, B. (2014, January 31). California drought prompts first-ever "zero water allocation." *Los Angeles Times*. Los Angeles, CA. Retrieved from http://articles.latimes.com/2014/jan/31/local/la-me-ln-california-drought-zero-water-allocation-20140131

Brant, J. a., Kwan, P., Daniel, U., & Valencia, R. (2010). Pilot-scale evaluation of chemical cleaning protocols for organic and biologically fouled microfiltration membranes. *Journal of Environmental Engineering, 136*, 542–553. doi:10.1061/(ASCE)EE.1943-7870.0000192.

CA DWR. (2015a). *Snow water equivalents (inches): February 6, 2015*. Sacramento, CA. Retrieved from http://cdec.water.ca.gov/cgi-progs/snow/DLYSWEQ

CA DWR. (2015b). *Sustainable groundwater management*. Sacramento, CA. Retrieved from http://www.water.ca.gov/groundwater/sgm/index.cfm

California Health & Safety Code §116551 (1998). Retrieved from http://www.oclaw.org/research/code/ca/HSC/116551./content.html#.VgxuK_m6dD8

Cayan, D. R., Das, T., Pierce, D. W., Barnett, T. P., Tyree, M., & Gershunov, A. (2010). Future dryness in the southwest US and the hydrology of the early 21st century drought. *Proceedings of the National Academy of Sciences, 107*, 21271–21276. doi:10.1073/pnas.0912391107.

CEC (2006). *Refining estimates of water-related energy use in California*. Sacramento, CA: California Energy Commission. Retrieved from: http://www.energy.ca.gov/pier/project_reports/CEC-500-2006-118.html

City of Los Angeles. (2008). The Water Conservation Plan of the City of Los Angeles. Los Angeles, CA. Retrieved from https://law.resource.org/pub/us/code/city/ca/LosAngeles/Municipal/chapter12.pdf

Cooley, H. & Gleick, P. H. (2011). Climate-proofing transboundary water agreements. *Hydrological Sciences Journal, 56*, 711–718. doi:10.1080/02626667.2011.576651.

Cousins, J. J., & Newell, J. P. (2015). A political–industrial ecology of water supply infrastructure for Los Angeles. *Geoforum, 58*, 38–50. doi:10.1016/j.geoforum.2014.10.011.

CWF (2014). *An evaluation of California groundwater management planning*. Sacramento, CA: California Water Foundation.

Dale, L. (2004). Electricity price and Southern California's water supply options. *Resources, Conservation and Recycling, 42*(4), 337–350. doi:10.1016/j.resconrec.2004.03.002.

Famiglietti, J. S., Lo, M., Ho, S. L., Bethune, J., Anderson, K. J., Syed, T. H., & Rodell, M. (2011). Satellites measure recent rates of groundwater depletion in California's Central Valley. *Geophysical Research Letters, 38*, 1–4. doi:10.1029/2010GL046442.

Gleick, P. H. (2008). Can California's water problems be solved? *Ecology Law Currents, 35*, 71–79.

Gleick, P. H. (2010). Roadmap for sustainable water resources in southwestern North America. *Proceedings of the National Academy of Sciences, 107*, 21300–21305. doi:10.1073/pnas.1005473107.

Griffin, D. & Anchukaitis, K. J. (2014). How unusual is the 2012–2014 California drought ? *Geophysical Research Letters, 41*, 9017–9023. doi:10.1002/2014GL062433.1.

Hering, J. G., Waite, T. D., Luthy, R. G., Drewes, J. E., & Sedlak, D. L. (2013). A changing framework for urban water systems. *Environmental Science & Technology, 47*, 10721–10726. doi:10.1021/es4007096.

Hughes, S., Pincetl, S., & Boone, C. (2013). Triple exposure: Regulatory, climatic, and political drivers of water management changes in the city of Los Angeles. *Cities, 32*, 51–59. doi:10.1016/j.cities.2013.02.007.

Hung, M.-F., & Chie, B.-T. (2012). Residential Water Use: Efficiency, Affordability, and Price Elasticity. *Water Resources Management, 27*(1), 275–291. doi:10.1007/s11269-012-0185-z.

Karl, T. R., Melillo, J. M., & Peterson, T. C. (Eds.). (2009). *Global Climate Change Impacts in the United States*. New York: Cambridge University Press. Retrieved from www.globalchange.gov/usimpacts

King, C. W., Stillwell, A. S., Twomey, K. M., & Webber, M. E. (2013). Coherence between water and energy policies. *Natural Resources Journal, 53*(1), 117–215.

Konikow, L. (2015). Long-term groundwater depletion in the United States. *Groundwater, 53*(1), 2–9. doi:10.1111/gwat.12306.

LA DWP (2010). *Urban water mangement plan*. Los Angeles, CA: Los Angeles Department of Water and Power.

LA DWP. (2015). LA DWP facts and figures. *LA Department of Water and Power Webpage*. Retrieved July 18, 2014, from https://www.ladwp.com/ladwp/faces/ladwp/aboutus/a-water/a-w-factandfigures?_adf.ctrl-state=hi9k84ta2_22&_afrLoop=1527136918451524

Lazarova, V., Oelker, G., & Won, W. (2012). Production of custom-made recycled water for various reuse purposes: Lessons learned from one of the world's largest recycling facilities. *Water Practice and Technology, 7*, 1-12. doi:10.2166/wpt.2012.052.

Leslie, B. J. (2014). Los Angeles, City of Water: The New York Times. December 6, 2014. Retrieved from: http://www.nytimes.com/2014/12/07/opinion/sunday/los-angeles-city-of-water.html

Lofman, D., Petersen, M., & Bower, A. (2002). Water, energy and environment nexus: The California experience. *International Journal of Water Resources Development, 18*(1), 73–85. doi:10.1080/07900620220121666.

Mehan, T. & Kline, I. (2012). Pricing as a demand- side management tool: Implications for water policy and governance. *Journal of American Water Works Association, 104*, 61–66.

Mo, W., Wang, R., & Zimmerman, J. B. (2014). Energy-water nexus analysis of enhanced water supply scenarios: A regional comparison of Tampa bay, Florida, and San Diego, California. *Environmental Science & Technology, 48*, 5883–5891. doi:10.1021/es405648x.

Plappally, A. K., & Lienhard, V. J. H. (2012). Energy requirements for water production, treatment, end use, reclamation, and disposal. *Renewable and Sustainable Energy Reviews, 16*(7), 4818–4848. doi:10.1016/j.rser.2012.05.022.

Pulido-Velazquez, M., Jenkins, M. W., & Lund, J. R. (2004). Economic values for conjunctive use and water banking in southern California. *Water Resources Research*, 40, n/a–n/a. doi:10.1029/2003WR002626.

Sanders, K. T. (2015). Critical review: Uncharted waters? The future of the electricity water nexus. *Environmental Science & Technology, 49*(1), 51–66. doi:10.1021/es504293b.

Sanders, K. T., King, C. W., Stillwell, A. S., & Webber, M. E. (2013). Clean energy and water: assessment of Mexico for improved water services and renewable energy. *Environment, Development and Sustainability, 15*, 1303–1321. doi:10.1007/s10668-013-9441-5.

Sanders, K. T. & Webber, M. E. (2012). Evaluating the energy consumed for water use in the United States. *Environmental Research Letters, 7*, 034034. doi:10.1088/1748-9326/7/3/034034.

State of California Water Code Sections 13560-13569 (2013). Retrieved from http://www.leginfo.ca.gov/cgi-bin/displaycode?section=wat&group=13001-14000&file=13560-13569.

Stillwell, A. S., Twomey, K. M., Osborne, R., Greene, D. M., Pedersen, D. W. P., & Webber, M. E. (2011). An integrated energy, carbon, water, and economic analysis of reclaimed water use in urban settings: a case study of Austin, Texas. *Journal of Water Reuse and Desalination, 1*, 208–223. doi:10.2166/wrd.2011.058.

Stroshane, T. (2008). Water and technological politics in California. *Capitalism Nature Socialism, 14*, 34–76. doi:10.1080/10455750308565523.

Wetterau, G. D., Chalmers, R. B., Liu, P., & Pearce, W. (2013). Advancing indirect potable reuse in California. *Water Practice and Technology, 8*, 275–285. doi:10.2166/wpt.2013.029.

Wilkinson, R. (2000). Methodology for analysis of the energy intensity of California's water systems, and an assessment of multiple potential benefits through integrated water–energy efficiency measures. Exploratory research project supported by Ernest Orlando Lawrence Berkeley Laboratory, Californa Institute for Energy Efficiency, Santa Barbara, CA.

Wilkinson, R. C. (2007). *Analysis of the energy intensity of water supplies for West Basin Municipal Water District*. Report for the West Basin Municiipal Water District, Carson, CA.

Quantifying and managing urban water-related energy use systemically: case study lessons from Australia

Steven John Kenway and Ka Leung Lam

School of Chemical Engineering, The University of Queensland, Brisbane, Australia

ABSTRACT

In this paper, three Australian case studies contribute to improved understanding of water-related energy quantification and management. A systems analysis of urban water in South East Queensland (Case Study 1) demonstrates the energy impact of water end use. In Melbourne (Case Study 2), water–energy interlinkages are explored within households. Finally, Case Study 3 in Sydney shows how abatement curves can help guide management action. Collectively, the case studies provide new information for least-cost solutions and simultaneous water and energy efficiency. The work highlights the need for frameworks to characterize and evaluate both the direct and indirect energy influences of urban water.

Introduction

Between 2005 and 2007, Australia was in the midst of the Millennium Drought, the period of least rainfall since 1788. Five major cities faced crisis-level water shortages. Headlines warned of "Armageddon" (ABC News, 2007). It was said that cities had to "desalinate or die" (Stolz, 2007). Climate change was widely held responsible (Bernstein et al., 2007; Howe, Jones, Maheepala, & Rhodes, 2005; WSAA, 2006). The crisis spawned significant urban water infrastructure investment to improve security. In major cities alone, a then record AUD 2.4 billion was spent in 2004–05, primarily for new water supply options, and this grew to over AUD 14 billion in 2008–09 (Kenway & Lant, 2012; WSAA, 2009).

Between 2008 and 2013, rainfall alleviated water shortages and investment pressures. However, a new challenge has emerged. The energy demand for water in Australian cities is anticipated to grow to 200–250% of 2007 levels by 2030, and the cost of energy embedded in the water to increase some 600% (Cook, Hall, & Gregory, 2012; Kenway, Priestley et al., 2008). This is largely due to the shift to energy-intensive infrastructure, but also attributed to population growth – from 12.1 million to 15.8 million people in the cities evaluated (Sydney, Melbourne, Brisbane, Perth and Adelaide), i.e. around 30%. An increase in energy use for urban water supplies is one problem, but combined with an expected rise in

Supplemental data for this article can be accessed at http://dx.doi.org/10.1080/07900627.2015.1132195.

electricity costs (AUD/MWh), the energy bill for urban water is anticipated to increase six-fold for most Australian cities (Cook et al., 2012). This now represents a significant business risk to the Australian water sector (Victorian Water Industry Association, 2011) and a cost risk to consumers.

To compound the challenge, Australia has now established a goal to reduce greenhouse gas (GHG) emissions, with "a trajectory range for emissions reductions of between 40 and 60% below 2000 levels by 2030" (Australian Government, 2014). If the water sector is to contribute proportionately, then energy use for urban water must be reduced by 75–85% from the projected 2030 levels (i.e. reduced to on-sixth or one-quarter of forecast 2030 levels). Alternatively, equivalent reductions would need to be made in the GHG emissions intensity of the fuel sources from which the water sector derives its energy. In Australia this is predominantly coal-fired power.

Given the magnitude of the urban water energy challenge, and the extreme situation through which Australia has travelled, this research focused on (1) understanding how water-related energy can be quantified and (2) the associated urban water resources management implications. In particular, tools which quantify energy use in various parts of urban water systems are considered. This article uses three case studies to illustrate how Australian water utilities have assessed the energy implications of urban water management in the post-drought period. A conceptual framework developed by the authors for analyzing urban water-related energy is used in this article to structure the discussion (Kenway et al., 2015).

Defining water-related energy

In this work, *water-related energy* is defined as representing the 'interconnections' or 'cause-and-effect' relationships between water and energy. When a change occurs in water use or water infrastructure (such as increased consumption or installation of new supply sources), changes in energy use will occur if the energy is 'related' to the water. For example, energy is used for producing and distributing drinking water, heating water for end use, and transferring and treating wastewater.

Water-related energy can be considered either *direct* or *indirect* from the perspective of water resource managers. For water utilities, for example, direct consequences lead to a shift in energy use of the utility. Indirect impacts of water on energy are more distant. For example, water conservation policy can influence household or industrial water use. While this may have a direct impact on utility energy use, it can also influence energy use within the household or industry – an indirect impact from the perspective of the utility.

Many water-related energy studies have been conducted on specific components of urban water, largely the energy use of water supply and/or wastewater systems (Lundie, Peters, & Beavis, 2004; Racoviceanu, Karney, Kennedy, & Colombo, 2007; Stokes & Horvath, 2009) and the energy implications of water management approaches such as water conservation and water loss management (Anand & Apul, 2011; Racoviceanu & Karney, 2010; Stokes, Horvath, & Sturm, 2013). Similarly, the energy impacts of alternative water supply options such as seawater desalination and interbasin water transfers have been addressed in the literature (Shrestha, Ahmad, Johnson, Shrestha, & Batista, 2011; Stokes & Horvath, 2006, 2009).

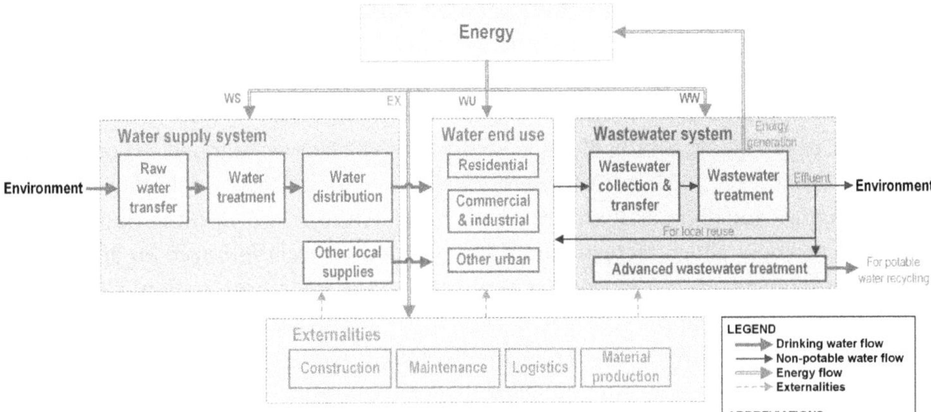

Tool	Case study	Component(s) considered for energy analysis
Systems modelling	1	WS (only for the centralised bulk water system)
Benchmarking	1	WS (only for the centralised bulk water system)
Life cycle analysis	1	WS, WW, EX
Material flow analysis	2	WU (only for the residential sector)
Marginal abatement cost curve	3	WS, WW

Figure 1. Overview of the four main components of the urban water system and its interaction with energy.

Case study analysis

In the literature, there is a general lack of articles reviewing the application of modelling results and tools for managing energy impacts of water (Kenway, Lant, Priestley, & Daniels, 2011c). This gap is partially addressed here by firstly outlining the major methods used to characterize the direct and indirect energy impacts of urban water. Secondly, three Australian case studies which use these tools to explore the associated water management implications are presented. By necessity, the case studies can only consider facets of the research projects which they describe; however, the online supplemental material (available at http://dx.doi.org/10.1080/07900627.2015.1132195) and detailed published journal articles and reports provide more information on the methodology, data and results associated with each of the projects described below.

Case Study 1 (South East Queensland) shows how the 'entire' energy influence of the urban water system can be characterized and evaluated, including both direct and indirect water-related energy impacts drawing on the analysis. Among other things, this case study demonstrates that water use for residential, industrial and commercial purposes is highly significant in the overall energy effect of urban water.

Case Study 2 (Melbourne) uses detailed analysis to explore water–energy linkages in the residential sector. This helps unpack the large energy effect which is indirectly influenced by water utilities.

Case Study 3 (Sydney) showcases how a least-cost tool has been used to guide energy management decisions, albeit with a focus on direct energy implications.

Conceptual framework and system definition

This article emphasizes the need to clearly define the boundaries of urban water systems. It expresses the importance of clarity such as whether the energy impacts being considered are directly within the water supply system, or indirectly, for example, in the water end use, wastewater or externalities, as presented in Figure 1. Approaches for quantifying water-related energy in each of these systems are documented more fully elsewhere (Kenway et al., 2015).

A system viewpoint is important for understanding water–energy links. Several researchers discuss the importance of understanding the 'system boundary' in order to make effective decisions (Decker, Elliott, Smith, Blake, & Rowland, 2000; Flower, Mitchell, & Codner, 2007; French & Geldermann, 2005; Satterthwaite, 2008). A system boundary definition is a critical first step in any modelling analysis (Sterman, 1991). Without knowing the boundary, it is impossible to know which factors should be included or excluded from an analysis. The boundary unequivocally influences decisions of the apparent 'best' option (Flower et al., 2007; French & Geldermann, 2005). This is particularly important in water analysis when the urban water system is not managed by a single entity.

The boundaries of the system in question will have considerable impact on the results of an analysis and the resultant management or policy decisions. A clear system boundary definition is important for meaningful decisions. A defined boundary helps identify limitations of information presented, and enables consistent comparison of results, through time or across different studies.

In this work, four major system components were defined for an urban water system (Figure 1) to quantify water-related energy use, and linked to the three case studies and the associated tools used. All four components have influences on energy, as follows:

- The water supply system consists of both the centralized water supply system managed by water utilities and any decentralized local supply sources (e.g. rainwater tanks, greywater reuse). For a typical main water supply, raw water is obtained from sources such as rivers, dams, aquifers, the sea, and effluent from wastewater treatment (in the case of direct potable water recycling). They are then treated into potable water and distributed to the urban water end users. Decentralized sources of water (e.g. rainwater tanks or bore water) can also have energy consequences, for example through their use of pumps.
- The wastewater system includes wastewater collection, transfer and treatment. The treated wastewater is normally discharged back to the environment. In some regions, treated wastewater is also used for nonpotable purposes such as landscape and industrial use. In some water-scarce regions, wastewater treatment effluent is further treated to a potable level for groundwater recharge. There are also energy consequences of nutrient discharges in wastewater and energy recovery (e.g. heat) or generation (e.g. via methane) in wastewater systems.
- Urban water end use represents all the water consumption activities in cities. The largest portion is usually for the residential sector. The use of hot water in residential end use is a significant part of water-related energy use in the urban water cycle. The influence of commercial and industrial water use on energy is also often considerable. Water–energy links in agriculture are also significant; however, they are not specifically addressed in this article.
- The externalities component is any other activities that contribute to the construction and operation of the three physical system components (water supply system, water end use and wastewater system). For example, it includes the energy use for the construction of treatment processes, and the energy use for the production and transport of chemicals used in water treatments.

Outline of available approaches and tools for quantifying and managing water-related energy

The five major approaches (and tools/types of tools) as described here are relevant to quantification of the water-related energy and served different purposes, given the definition of the system adopted. These tools were applied in the three case studies (see the list in Figure 1). A brief description is given below, with further detail in the online supplemental material. The main tools used for characterizing water-related energy considered in this research include the following.

- *Systems modelling* is typically water balance or quality modelling of a particular component of urban water (e.g. distribution or treatment). Energy considerations are typically added as a secondary stage. This approach was used in Case Study 1 to establish the direct energy use of the bulk water supply authority (Seqwater).
- *Benchmarking* approaches quantify the performance of a specific-system component relative to a comparison group. Energy-efficiency benchmarking in water is relatively new compared with other benchmarking efforts. This approach was used in Case Study 1 to compare the energy performance of different drinking water treatment plants (WTPs) operated by Seqwater.
- *Life cycle analysis* is a systematic approach for modelling the exchanges of resources between a product or service and the environment, from cradle to grave. For water systems and energy, this has mainly focused on the 'upstream' (supply chain) energy demands of various materials (such as chemicals, concrete and steel). This approach was used in Case Study 1 to quantify the energy embodied in chemicals used in potable water treatment and wastewater treatment.
- *Material flow analysis* has been used for diverse and detailed environmental problem-solving applications, typically focused on the flow of a particular material (e.g. copper or zinc). More recently, it has been used to quantify water–energy links within households associated with water use. This approach was used in Case Study 2 to quantify water-related energy of the various subcomponent systems (e.g. showers, baths, clothes washer).
- *A marginal abatement cost curve* tool presents the costs of implementing abatement activities in a graphical format that allows least-cost abatement options to be easily determined. As a tool it helps compare management or infrastructure options. Typically, creation and assessment of each option requires more detailed tools such as those described above. This approach was used in Case Study 3 to compare various water management options on cost per unit of GHG emissions reduction.

Case Study 1: Systemic energy efficiency in the urban water system in South East Queensland

Background

During the Millennium Drought, large-scale water demand management initiatives were promoted throughout Australia, including South East Queensland (SEQ). Significant infrastructure was also constructed, including a desalination plant (125 ML per day), an indirect potable water recycling system (232 ML per day) and a bulk water supply network (Queensland Water

Commission, 2010). Seqwater (the regional bulk water authority, http://www.seqwater.com.au) now manages these assets as well as associated catchments and WTPs. Water retailing and distribution are handled by local government–owned retail entities.

Since around 2010, a trend towards using more energy-intensive water supply options, a heavy reliance on costly grid electricity, and the introduction of the carbon tax (though it was later abolished) convinced Seqwater to conduct a project to (1) explore the extent to which energy and GHG emissions were influenced by urban water; (2) understand related costs; and (3) understand management approaches for efficiency (Kenway et al., 2015).

Approach adopted and tools used

System modelling (Lam, Kenway, & Lant, 2015) and life cycle analysis (Lane, de Haas, & Lant, 2015) were the primary approaches used to quantify water-related energy in the SEQ supply system (Kenway et al., 2015). To evaluate the key system components of the SEQ supply system, an energy benchmarking analysis of Seqwater's drinking WTPs was conducted. In the analysis, electricity use by the WTPs was (1) separated, using regression analysis where possible, into flow-independent use (for treatment) and flow-dependent use (for raw water and effluent pumping), and (2) compared across the WTPs.

To evaluate the bulk water supply system, a system modelling approach (Bartos & Chester, 2014; Baynes, West, Vitkovsky, & Hall, 2009; Shrestha et al., 2011) was used to quantify and project the energy use of the SEQ bulk water supply system. A long-term water balance optimization model was used to generate the future water balance, which was then used for energy quantification based on the compiled energy-intensity data of key assets (e.g. all major WTPs, desalination and water recycling facilities). This allowed the modelling and quantification of the energy use of the bulk water supply system under various scenarios (such as the extent of the use of desalination operations, future demand and future catchment inflow scenarios).

Life cycle analysis was undertaken for the Gold Coast (where some 700,000 people live, a subset of the SEQ system) considering the life cycle impacts (i.e. including energy use) for the provision of water supply and wastewater management services in 2009 (Lane et al., 2015). The analysis was conducted to improve the understanding of indirect energy use in externalities, particularly those related to the supply chain of chemicals for water treatment. Due to high data demands, the study focused on the Gold Coast's system only (i.e. a large and arguably representative subsection of the full SEQ water system).

Guided by detailed mathematical material flow analysis (MMFA) modelling (conducted in Case Study 2), a city-scale water-energy balance model was used to quantify water-related energy in the residential sector. Water-related energy in the commercial and industrial sectors was estimated guided by state-wide data, and energy use for wastewater treatment by communication with available utilities (see Kenway et al., 2015, for details).

In order to compare the use of different forms of energy (electricity, natural gas, etc.), it was necessary to convert energy use into primary energy equivalents. This was based on an assumption of three units of thermal energy being required to produce one unit of electrical energy (Gleick & Cooley, 2009; Kenway, Lant, & Priestley, 2011b). Natural gas was assumed to be purchased as primary energy.

Table 1. Electricity use breakdown for major drinking water treatment plants in South East Queensland (Kenway, Lane, et al., 2014).

<table>
<tr><th rowspan="3">Drinking water-treatment plant</th><th>Flow range</th><th colspan="2">Flow-independent</th><th colspan="3">Flow-dependent</th><th rowspan="2">Total</th></tr>
<tr><th></th><th colspan="2">Treatment</th><th>Raw water pumping</th><th>Treatment</th><th>Product water discharge</th></tr>
<tr><th>ML per day</th><th>MWh per day(% of total)</th><th>kWh/ML</th><th>kWh/ML</th><th>kWh/ML</th><th>kWh/ML</th><th>kWh/ML</th></tr>
<tr><td>1</td><td>230–360</td><td>11.3 (9%)</td><td>38</td><td>384</td><td>8</td><td>17</td><td>447</td></tr>
<tr><td>2</td><td>70–80</td><td>1.4 (26%)</td><td>19</td><td>0</td><td>44</td><td>11</td><td>73</td></tr>
<tr><td>3</td><td>20–110</td><td>2.4 (10%)</td><td>30</td><td colspan="2">75</td><td>184</td><td>289</td></tr>
<tr><td>4</td><td>60–140</td><td>1.1 (9%)</td><td>9</td><td>37</td><td>56</td><td>0</td><td>102</td></tr>
<tr><td>5</td><td>1.0–1.4</td><td>0.1 (17%)</td><td>101</td><td colspan="2">502</td><td>0</td><td>603</td></tr>
<tr><td>6</td><td>1.3–2.0</td><td>0.3 (21%)</td><td>180</td><td>533</td><td colspan="2">140</td><td>853</td></tr>
<tr><td>7</td><td>6–9</td><td>0.4 (7%)</td><td>58</td><td>483</td><td colspan="2">254</td><td>795</td></tr>
<tr><td>8</td><td>40–70</td><td>1.3 (19%)</td><td>26</td><td>102</td><td>8</td><td>0</td><td>135</td></tr>
<tr><td>9</td><td>15–29</td><td>0.0 (0%)</td><td>0</td><td colspan="2">662</td><td>0</td><td>662</td></tr>
<tr><td>10</td><td>17–22</td><td>1.3 (27%)</td><td>71</td><td>172</td><td colspan="2">21</td><td>264</td></tr>
</table>

Note: Energy intensity values presented mid-way between columns indicate that energy use could not be partitioned between these components of the water cycle.

Outcome and implications for water management

The benchmarking analysis of the major WTPs in SEQ confirms the expectation that pumping-related power demand (i.e. in the form of raw water pumping and product water discharge) dominates the overall WTP-related electricity use. Using (1) the data in Table 1 for four of the largest plants (nos. 1–4 in Table 1) where more detailed breakdown is available and (2) the production flows of the plants from 2010 to 2012 (279, 74, 73 and 112 ML per day, respectively), a total flow-weighted estimate indicates that Seqwater's overall energy use profile was approximately 9% flow-independent treatment, 70% flow-dependent raw water pumping, 7% flow-dependent treatment and 14% flow-dependent product water pumping. Despite the major influence of pumping operations, a substantial component (up to ~25% of individual plants) of the electricity use at many of the WTPs (for both small and large plants) appears to be independent of flow rate. This benchmarking analysis also unveiled that there is no sub-metering for most of the plants (which would help optimize daily operations) to allow a consistent comparison of energy use across all WTPs.

The following conclusions were observed with regard to primary energy use throughout the overall SEQ urban water system (not including all the embedded energy) in 2011–12 (see also Table A2 in the online supplemental material and Kenway et al., 2015, for details):

- Water end use is responsible for over 89% of the primary energy use of the total SEQ urban water system (61% by residential water users and 28% by commercial/industrial users). Wastewater treatment used approximately 6% and water provision used around 4.5% (3.5% for Seqwater) of system-wide energy.
- Collectively, water-related energy in the SEQ urban water system accounted for some 13% of the electricity use in the region, plus 18% of the total natural gas use, plus 3.3% of all 'other' energy consumed in SEQ.
- Within the fraction of energy use influenced by Seqwater (i.e. 3.5% of the total system), drinking WTPs were responsible for 60% of total energy use, followed by desalination

(20% for limited production) and the bulk water supply network and water recycling system (approximately 10% each). The energy benchmarking analysis identified the major energy-consuming asset to be focused on raw water pumping, and often concentrated at individual water treatment assets. For instance, WTP 1 in Table 1 used approximately 70% of total electricity consumption by all WTPs in SEQ to produce around 40% of the mains water supply.

- While this study did not quantify virtual water flow through SEQ (arguably an externality influenced by urban water strategies), it is expected to be substantial, because virtual water flows through similar Australian cities are approximately eight times the direct water flows, based on analysis of Melbourne (Kenway, Turner, Cook, & Baynes, 2008).
- The life cycle analysis for the Gold Coast urban water system (Lane et al., 2015), which was conducted as a component of the overall investigation of SEQ, shows that the embedded energy in chemicals used for potable water treatment is similar to the direct energy use for water provision. On the other hand, the embedded energy in chemicals for wastewater treatment is relatively small compared to the direct energy use for wastewater treatment.

Implications for water resources management

This work has generated new information which helps understand and manage energy and related GHG emissions through the SEQ water system. This has implications relevant to other regions. For example:

- Knowledge of the dominance of water end use on the energy impact of the urban water cycle arguably should lead to more inclusion of energy costs and benefits in studies which consider the costs and benefits of alternative water use patterns. For example, while water and wastewater utilities in SEQ spent some AUD 50 million for electricity in 2011–12 (Table A2 in the online supplemental material), the cost of energy expenditure *influenced* by water use is over 20 times as much (~AUD 1,100 million per year). This expenditure is primarily associated with residential, industrial and commercial water use.
- A major gap identified is that energy consumption statistics are not available in SEQ other than for total electricity delivered by the electricity distributor. It would be very helpful if improved energy data (including use by sector and fuel type) were available in SEQ. Such data is viewed as highly relevant for managing the future energy efficiency of the regional water system.
- The work provides insight into the expected efficiency (or potential inefficiency) of each WTP, as well as which plants require further analysis for further partitioning of electricity consumption data. This could be used to help prioritize understanding and management of different WTPs and related steps including sub-metering, particularly in those plants where a breakdown of the energy consumption was not possible.

Case Study 2: Residential water-energy analysis in Melbourne

Background

The significance of water end use in the energy consumption of the urban water cycle is made clear in Case Study 1: water end use accounted for 89% of energy use in the SEQ urban water system. Water end use is also broadly emphasized as important in the water–energy nexus by a range of authors (Arpke & Hutzler, 2006; Cheng, 2002; Klein, Krebs, Hall, O'Brien, & Blevins, 2005). Because the energy influence of water use is so large, particularly in the residential sector, this case study focuses on that component.

To efficiently manage water-related energy consumption, a detailed understanding of water–energy links is critical. Robust knowledge of linkages in households is a good place to start because the connections are significant, but also because the sector is responsive to government policy, for example via incentives, pricing programs, technology deployment and information provision (Kenway & Tjandraatmadja, 2009). The residential sector has been identified as an area in which to prioritize water–energy interactions research (Kenway et al., 2011c) due to strong interlinkages, and generally greater data availability compared to industrial and commercial water use (which is often more sensitive in nature and site-specific).

In order to progress an understanding of water–energy links within households, a research project was initiated, focusing on Melbourne households, in collaboration with the responsible water and energy utilities and government agencies, including Yarra Valley Water, City West Water, South East Water, Jemena Electricity, Envestra (gas) and the Victorian state government (Kenway, Binks, et al., 2014). The research aimed to elucidate and quantify water–energy–carbon links in households, as well as groups of households. It considered how actions and policies taken by the water industry (e.g. water demand management programs, water pricing strategies and even changes to water asset management, which influences the temperature of water delivered) can influence energy use in private households. In this way, the work goes beyond the boundary of traditional water utility energy use analysis, which typically assesses the energy implications directly connected with the utilities themselves, such as the energy demands of the water and wastewater infrastructure and assets.

Water end use research (Beal, Stewart, & Bertone, 2012; Roberts, 2012) has enabled Australian water utilities to achieve high levels of water conservation by informing both government and customers where to target change. Industry has responded through innovation to improve appliance efficiencies. High uptake of water-efficient measures such as low-flow shower nozzles, dual-flush toilets, and front-loading clothes washers has achieved considerable water savings for both utilities and customers.

Detailed assessment is important because some water efficiency measures can result in unintended outcomes. For example, changing from a top-loading clothes washer plumbed for gas-heated water, to a water-efficient front-loading machine plumbed for cold water only, with internal electric heating, could substantially increase GHG emissions from the clothes washer, even if energy use decreases (Kenway, Scheidegger, Larsen, Lant, & Bader, 2013). This can happen if the new machine has only a "cold water" connection, as opposed to being able to receive flows of both hot and cold water. Where new efficient machines have only a cold water connection, it is no longer possible for the machine to accept hot water delivered by the hot water system; as a result, it then relies on electrical heating within the clothes washer to warm the water. If the household had a solar or natural gas hot water system which previously supplied the machine, and if the electrical energy is provided by coal-fired electricity, then the forced change in energy used to heat the water can increase GHG emissions. The increase in GHG emissions occurs because electricity produced from coal-fired plants in Australia (1.04 kg CO_2-e/kWh, carbon dioxide equivalent per kilowatt hour) has more than five times the GHG emissions per unit of energy delivered, compared to natural gas (0.197 kg CO_2-e/kWh – Commonwealth of Australia, 2008) and solar hot water systems. The importance of plumbing configurations for the GHG emissions impact of water conservation (via clothes washers) highlights the need for detailed understanding of water–energy–GHG emission linkages in order to avoid unintended outcomes such as increasing GHG emissions through water conservation programs. Development of methodologies to quantify water-related energy has also been identified as a priority for progressing management of water-related energy in cities, and a precursor for establishing management goals or targets (Kenway, McMahon, Elmer, Conrad, & Rosenblum, 2013).

Approach adopted and tools used

Seven households were surveyed (five in Melbourne, two in Brisbane) using behavioural interviews and technical audits of water fittings, fixtures and appliances (Binks, Kenway, Lant, & Pamminger, 2014). Five years of billing records for water, energy, and natural gas use were collected. Energy and water flow meters (Amphiro) were installed in showers, together with log-sheets to collect data on frequency, duration, temperature, and flow rate of shower events (due to their substantial impact on water and energy usage). Using data from all these sources, an MMFA model (Kenway et al., 2013) was populated with data for 139 parameters of relevance to sub-household flows of water and energy. The model was used to generate result data-sets (water, energy, GHG, and cost flows at subsystem level) and create a detailed report for each individual household (Binks & Kenway, 2014). The detailed method is described by Binks et al. (2014).

Outcome and implications for water management

The results (Figure 2) indicate that water-related energy usage ranges from 5 to 20 kWh/hh·d (kilowatt hour per household per day), or 2.5–5.0 kWh per person per day (Kenway, Binks, et al., 2014). Of this, 2.0–7.2 kWh/hh·d was related to shower water use. Collectively, water-related energy represented 15–50% of total direct energy use in the households studied, noting that this excludes energy needed to supply water to the households or to remove wastewater from the households, e.g. in the wastewater treatment system. For perspective, 10 kWh per day is equivalent to operating 37 high-efficiency (11 W) light bulbs 24 hours per day.

The MMFA was used to conduct local sensitivity analyses to identify key factors which, if changed in small amounts, can have the greatest influence on water-related energy usage. Preliminary analysis indicated that key factors include: (1) the temperature of showers for adults (noting that only two households had children and the energy use of showers or baths for children was lower than for adults due to cooler water temperatures and shorter durations of water use); (2) the duration, flow rate and frequency of showering per adult; (3) the number of adults per household; (4) the supply temperature of cold water entering the house; and (5) the efficiency of the hot water system.

Preliminary sensitivity analysis indicated that small changes to water management within – or outside – the household could have a surprisingly large impact on energy use for the consumer. They could also influence energy use for water and wastewater treatment and transport. For example, a 10% reduction in the frequency, duration, or flow rate of water for showers would reduce household energy use by approximately 0.3 kWh/hh·d in some households. Preliminary analysis indicates that a 10% change in supplied water temperature (approximately 2 °C) could influence household energy use by around 0.5–1.0 kWh/hh·d

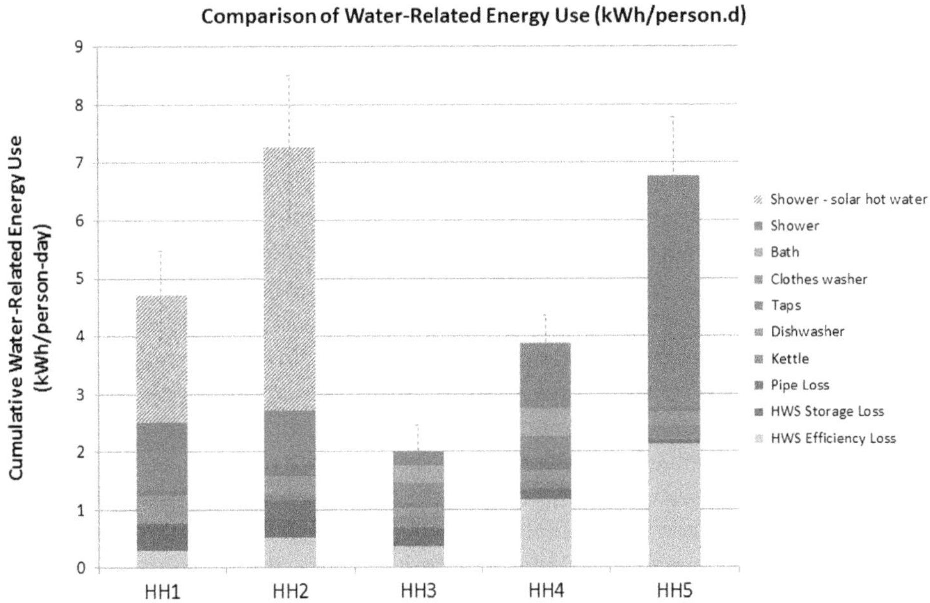

Figure 2. Water-related energy in five Melbourne households. (Source: Data described in Binks & Kenway, 2014, and Kenway, Binks, et al., 2014.)

(noting a wide range in the individual households studied). Consequently, accurate characterization of cold water temperature is an important factor in quantifying water-related energy consumption. For the five Melbourne households studied, reducing by 1 kWh/hh·d use over an entire year would reduce household costs by AUD 70–125 for electricity or AUD 16–30 for natural gas.

Guided by the detailed modelling of individual households, spatial analysis of key factors was undertaken. For example, analysis of the supply temperature at over 40,000 locations in the Melbourne water network has been used as an input to the MMFA modelling (Bors & Kenway, 2014). This found significant temporal variability and surprisingly high spatial variability: supply temperature can vary as much as 8 °C (from 15 to 23 °C) within an approximate 3 km zone in a single month. Considering the sensitivity analysis, such variability could influence energy use by on the order of 2–4 kWh/hh·d. In moving forward towards a full city-scale analysis, a number of challenges have emerged. These include the adequacy of integrated data-sets, not across only the water and energy sectors but also in related data, such as census data from the Australian Bureau of Statistics, appliance stock data, etc.

The work to date has helped identify the water management options which have the greatest scope for influencing household energy consumption. Such options include for example technologies or behaviour-management programs which influence shower duration, frequency, flow rate or temperature. This will help design energy-effective water conservation policies and help avoid unintended outcomes such as increased GHG emissions attributable to water efficiency strategies (e.g. from water-efficient but electricity-intensive appliances). Such knowledge also helps improve policy beyond 'average' results and shift towards policy tailored to individual household configurations. Quantification of energy impacts of water will also enable utility energy management strategies (e.g. on an AUD/kWh basis) to be compared to options which mitigate household energy use. This would enable least-cost options to be identified considering alternatives within and beyond utility boundaries. This information is perceived as critical to enabling utilities to invest in water conservation in the ways which have the strongest benefits for within-household energy use. Such knowledge is also expected to lead to a shift towards greater collaboration in water and energy retail services (for example in combined metering, technology rebates and information provision), some of which is already observed in California (EcoNorthwest, 2011).

The work also gives insight into water asset design and management implications. For example, the analysis of the variability of water temperature suggests that an 8 °C difference in water temperature could change energy use by up to ~4 Wh/hh·d energy use (Kenway, Binks, et al., 2014). Energy density mapping of the water and wastewater network indicates that most households in Melbourne require 1–2 kWh/m^3 for provision of water and wastewater services (Saliba & Gan, 2006). Based on 0.5 m^3/hh·d water use, the energy for delivery of the water/wastewater service in the vicinity of the households studied is on the order of 0.5–1.0 kWh per day per household. This indicates that existing observed variation of water temperature could be influencing up to three times the energy consumption in households, compared to the energy needed for the water and servicing of those households. Consequently, further understanding of the various asset management strategies will identify energy-saving opportunities. For example, it is possible that alternative design and management of water infrastructure could substantially impact household energy use and costs.

A wide and integrated suite of programs and measures, including review of government policy instruments, is likely to be needed to progress change. A stakeholder workshop conducted for the project (Kenway, 2015) identified five areas warranting attention:

- Integrated water and energy metering
- Improved connection of water and energy efficiency labelling and assessment
- Establishment of state-wide water and GHG emissions targets
- Targeting water and energy programs to benefit specific customer types
- Further consideration of water and energy utility operation mission statements and goals.

The workshop also identified that, in order to progress improved management of water–energy interlinkages in households, further consideration needs to be given to a better understanding of the influence and role of institutional frameworks, governance arrangements, and financial incentive models for water and energy management in cities. Mechanisms for increasing the connection between service provision and revenue need to be strengthened. Likewise, decoupling revenue from volumetric sales of water and energy (Lesh, 2009) would provide alternatives to water utilities for their continued dependency on high throughput (sales) of water and energy.

While this case study focused on understanding residential households, the work also has implications for understanding water-related energy in commercial and industrial premises. For example, many commercial premises such as accommodation services, sporting facilities and prisons also have a range of water-using functions (such as showers) which have a large influence on energy use at the premises. Similarly, the strong influence of the temperature of cold water on household energy use, coupled with its high spatial variability, indicates

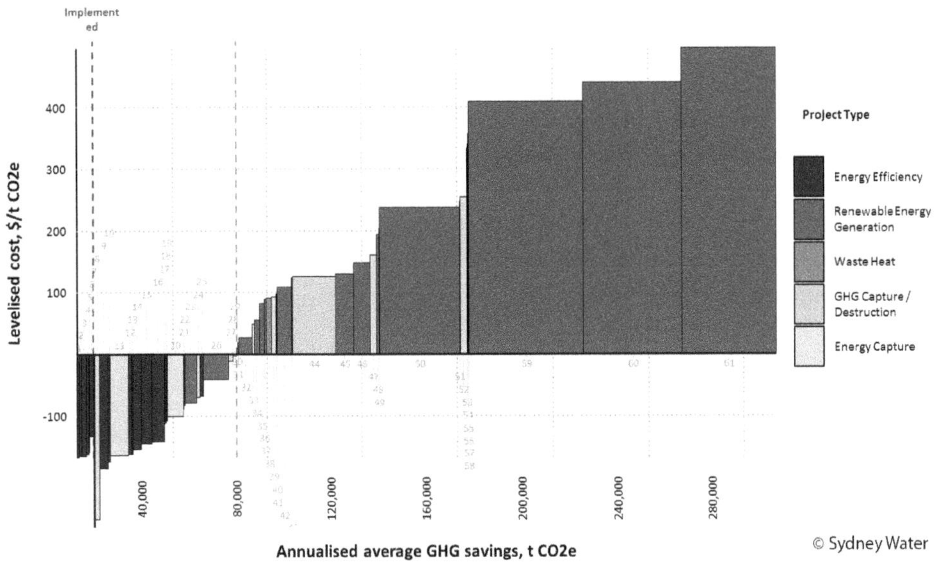

Figure 3. An example of a marginal abatement cost curve (Sydney Water, 2012). Note that the appropriateness of any particular solution is highly system-specific.

that further knowledge of this effect could help with understanding the energy influence of water use in different commercial and industrial operations.

Case Study 3: Least-cost carbon in water and wastewater services in Sydney

Background

Sydney Water provides drinking water, wastewater treatment, recycled water and stormwater services in Sydney, the Illawarra and the Blue Mountains. During the Millennium Drought, mandatory restrictions on water use were enforced. As in SEQ, a desalination plant was built in Sydney in response to the water crisis. The drought resulted in an increased use of alternative water sources such as recycled water, tank water and desalinated water. Sydney Water determined that there is a need for a tool to support decision making on GHG emissions abatement, and therefore developed a Cost of Carbon Abatement (CCA) tool (WSAA, 2012) and used it to prioritize GHG abatement options.

Tool used

The purpose of the CCA tool is to compare the cost-effectiveness of different GHG abatement options. The use of the tool involves defining abatement opportunity, defining generic inputs (e.g. common assumptions applied to all opportunities, scenario assumptions) and populating it with detailed data (e.g. utility-specific electricity use; carbon price projections; discount rates; capital, operating and maintenance expenditure; mandatory and voluntary carbon costs; risk assessment). The results are presented as a marginal abatement cost curve. While developing the curve is a data-intensive process, the curve allows least-cost abatement opportunities to be easily determined for any chosen GHG abatement target. Figure 3 shows an example curve, along with generic solutions, from energy efficiency (such as newer or variable-speed pumps) to energy capture (such as harvesting chemical energy within biogas or sludge). For example, an option that has a negative levelized cost is a cost-saving opportunity that can pay back within the time period studied (e.g. 20 years). In this example, most of these cost-saving options are energy-efficiency options. For this system, large-scale renewable energy generation options offer greater abatement potential, but they have a higher abatement cost. It is critical to note, though, that the solutions are highly system-specific (due for example to differences in local topography, infrastructure and operational conditions) and cannot be generalized.

Outcome and implications for water management

In 2010, Sydney Water conducted a project to apply the CCA tool to support strategic decision making on energy management and GHG emissions abatement (Sydney Water, 2010). In the project, 111 abatement opportunities were initially identified and evaluated in preliminary screening to ensure that the options were compatible with Sydney Water's operational constraints (WSAA, 2012). The 90 compatible opportunities went through a preliminary analysis to assess their net present values per unit of GHG abated, for prioritization. In the end, 37 opportunities were analyzed in detail using the CCA tool, which was populated with detailed financial data. The tool then generated a marginal abatement cost curve for each future scenario defined (e.g. changes to electricity and carbon prices). As a result of the analysis, Sydney Water has invested annually in least-cost energy-efficiency projects such as equipment replacement and refurbishment, lighting control, process improvements, pump curve optimization and ventilation control, and in increasing renewable energy generation such as cogeneration using biogas from digesters and mini-hydro power generation. In addition, the project allowed Sydney Water to identify some research opportunities, such as aeration system optimization, biogas process optimization and microbial fuel cells.

After the project, the CAA tool was modified to be more generic, for use by other water utilities (WSAA, 2012). As of 2014, the CCA tool has been licensed to 19 water utilities around Australia (Sydney Water, 2014), where they currently use the tool to manage direct energy use and GHG emissions.

Discussion

Implications for utilities

Case Study 1 clearly demonstrates the significance of indirect water-related energy, particularly the large quantities of energy influenced by the use of water for residential, industrial and commercial purposes. As water utilities play an active role in both supply and demand management of water, there is a strong case for utilities to take a much more active role in the management of the energy influences of water use.

Typically, water utilities manage direct energy use and cost with asset management strategies such as pump station upgrades or changes to treatment processes. These are quite different from the management options available for residential water (and energy) end use management as highlighted in Case Study 2. Given that far more energy is connected to water use than to water supply or wastewater disposal (as shown in Case Study 1), measures which influence the energy consumption of water use appear to have far greater scope than more traditional 'within utility' measures.

Importantly, however, to date the CCA (least cost) tool highlighted in Case Study 3 has not been used to compare the direct and indirect energy influence of different management options.

Implications for quantifying water-related energy

This work pays attention to quantification of direct and indirect water-related energy from the perspective of water utilities. The adequacy of current frameworks hinges significantly on which subsystems of 'the urban water system' are assessed. A current barrier to management of water-related energy is the lack of adequately developed guidance for this. For example, GHG emission accounts and management are guided by international (e.g. Greenhouse Gas Protocol, http://www.ghgprotocol.org/) and Australian protocols; however, comparable protocols (or analysis methods) do not yet exist for the quantification of water-related energy. An international framework for characterizing and evaluating both the direct and indirect influence of urban water-related energy would be of significant value.

In addition to an analytical framework, a clear conceptual framework is also needed to describe the direct and indirect influence of urban water on energy. This is important because conceptual frameworks have considerable impact on the design and use of methods and tools for evaluation of water–energy intersections.

A major challenge for our cities of the future is to reduce the inputs and outputs of water, energy, materials and food, but simultaneously increase human well-being and ecosystem health. 'Urban metabolism' is a conceptual framework which can clearly delineate the boundary of the 'city system' and consequently improve the quantification of urban performance (Kenway, 2013; Kenway, Gregory, & McMahon, 2011a; Kenway, Pamminger, & Lant, 2013; Pamminger & Kenway, 2008). Without good frameworks it is more likely that water and energy problems will continue to be moved from one domain to another, rather than being truly solved.

Conclusions

This work used three Australian case studies to explore the direct and wider indirect influences of urban water systems on energy use.

In Case Study 1, a systems view of urban water compared the water-related energy use of system components, including water supply, wastewater, water use and related externalities. The work demonstrated the significance of water use, which has a greater energy use impact than water utilities for water supply and wastewater services, even with the uptake of desalination in Australia in recent years. Although a relatively minor component of system-wide energy use, the energy demand for the bulk water supply system was still significant (410 gigawatt-hours of thermal energy) in 2011–12. Detailed systemic quantification of water-related energy is necessary for management to make sound judgement on appropriate strategies to enable cost-effective management of energy demands of urban water.

In Case Study 2, the strong 'energy for water' linkage in water use was examined in detail in households. The results indicate that water-related energy in households is particularly influenced by functions such as hot water for showers and baths, hot water system efficiency and clothes washer type. The analysis also demonstrated how the management of the water asset (the supply system) can influence household energy consumption, via the temperature of delivered water. The analysis provides new information for water utilities interested in finding least-cost solutions to rising water and energy costs. Such information is argued as relevant to utilities providing services to help customers achieve combined water and energy efficiency, and to design efficient overall water and energy systems, including the design of water supply assets.

In Case Study 3, it was shown how water utilities can use a management tool (the CCA tool) to compare, and prioritize for action and investment, various management options.

These insights are relevant to other cities interested in pathways towards greater energy efficiency for the 'water system'. The messages are important, not only in the provision of urban water or wastewater services, but also for overall urban energy efficiency.

Historically, the lack of quantitative information regarding direct and indirect links between water and energy has constrained the motivation for management. However, a range of new tools are emerging that improve quantification of water-related energy. While the progress is encouraging, to genuinely understand the energy influence of water in cities and infrastructure systems there remains a need to significantly develop analytical tools and related data-sets. There is also a need to refine the conceptual, analytical and theoretical frameworks surrounding the knowledge and management of water and its influence on energy.

Acknowledgement

The authors thank the Water-Energy-Carbon Research Group members for their work on related projects. In particular, we thank Dr Joe Lane for leading the water treatment plant analysis, Amanda Binks for surveying and modelling Melbourne households, and Dr Marguerite Renouf for review of an earlier draft of this paper.

Disclosure statement

No potential conflict of interest was reported by the authors.

Funding

The authors would like to thank the various research sponsors, including the Australian Research Council (LP 120200745), Smart Water Research Fund (and its component utilities), Seqwater, CRC for Water Sensitive Cities, and the Water Research Foundation.

References

ABC News. (2007). *Beattie scraps water poll amid 'Armageddon' situation*. Brisbane: Australian Broadcasting Corporation.

Anand, C., & Apul, D. S. (2011). Economic and environmental analysis of standard, high efficiency, rainwater flushed, and composting toilets. *Journal of Environmental Management, 92*, 419–428.

Arpke, A., & Hutzler, N. (2006). Domestic water use in the United States. A life-cycle approach. *Journal of Industrial Ecology, 10*, 169–183.

Australian Government. (2014). *Reducing Australia's greenhouse gas emissions – Targets and progress review*. Final Report. Canberra. Retrieved from Climate Change Authority website: http://www.

climatechangeauthority.gov.au/files/files/Target-Progress-Review/Targets%20and%20Progress%20 Review%20Final%20Report.pdf.

Bartos, M. D., & Chester, M. V. (2014). The conservation Nexus: Valuing interdependent water and energy savings in Arizona. *Environmental Science and Technology, 48*, 2139–2149.

Baynes, T., West, J., Vitkovsky, J., & Hall, M. (2009). *A framework for modelling the energy and greenhouse implications of water demand and supply scenarios*. Retrieved from Urban Water Security Research Alliance Technical Report No:. 15.

Beal, C. D., Stewart, R. A., & Bertone, E. (2012). *Implications of resource-efficient technology on peak water demand and water-related energy demand*. Urban Water Security Research Alliance Science Forum 2012, Brisbane.

Bernstein, L., Bosch, P., Canziani, O., Chen, Z., Christ, R., & Davidson, O. (2007). Climate change 2007: Synthesis report. Summary for policymakers. *Lima; CONAM, 2006, 71 p.(Serie Indicadores Ambientales, 10)*.

Binks, A., & Kenway, S. J. (2014). *Characterisation of water-related energy in households 1-5. Summary of analysis outcomes for five households in Melbourne, Australia*. Melbourne: Smart Water Fund. Retrieved from Clearwater website: http://clearwater.asn.au/resource-library/smart-water-fund-projects/water-energy-carbon-links-in-households-and-cities-a-new-paradigm.php.

Binks, A., Kenway, S.J., Lant, P., & Pamminger, F. (2014). Detailed characterisation of water-related energy use in households. In Australian Water Association (Ed.), *Ozwater 2014*. Brisbane: Australian Water Association.

Bors, J., & Kenway, S. J. (2014). *Water temperature in Melbourne and implications for household energy use*. Melbourne: Smart Water Fund. Retrieved from http://clearwater.asn.au//user-data/research-projects/swf-files/9tr1---001-grace-2014-cold-water-temperature-in-melbourne-1994-2013-final.pdf.

Cheng, C.-L. (2002). Study of the inter-relationship between water use and energy conservation for a building. *Energy and Buildings, 34*, 261–266.

Commonwealth of Australia. (2008). *National Greenhouse Accounts (NGA) Factors*. Canberra, Australia: Australian Government Department of Climate Change.

Cook, S., Hall, M., & Gregory, A. (2012). *Energy use in the provision and consumption of urban water in Australia: An update. A report prepared for the water services association of Australia*. Canberra.

Decker, E. H., Elliott, S., Smith, F. A., Blake, D. R., & Rowland, F. S. (2000). Energy and material flow through the urban ecosystem. *Annual Review of Energy and the Environment, 25*, 685–740.

EcoNorthwest. (2011). *Embedded energy in water pilot programs impact evaluation final report*. San Francisco. Retrieved from Caifornia Public Utilities Commission website: http://www.energydataweb.com/cpucFiles/33/FinalEmbeddedEnergyPilotEMVReport_1.pdf.

Flower, D. J. M., Mitchell, V. G., & Codner, G. P. (2007). *The potential of water demand management strategies to reduce the greenhouse gas emissions associated with urban water systems*. 1st Conference on Sustainable Urban Water Management & 9th Conference on Computing and Control in the Water Industry, Leicester, UK.

French, S., & Geldermann, J. (2005). The varied contexts of environmental decision problems and their implications for decision support. *Environmental Science & Policy, 8*, 378–391.

Gleick, P., & Cooley, H. (2009). Energy implications of bottled water. *Environmental Research Letters, 4*.

Howe, C., Jones, R., Maheepala, S., & Rhodes, B. (2005). *Melbourne water climate change study: Implications of potential climate change for Melbourne's water resources* Canberra: CSIRO.

Kenway, S. J. (2013). *The water-energy Nexus and urban metabolism - Connections in cities*. Brisbane: Urban Water Security Research Alliance.

Kenway, S. J. (2015). *Stakeholder Workshop 2: Outcomes Statement. Clearwater*. Retrieved from Clearwater: http://clearwater.asn.au/resource-library/smart-water-fund-projects/water-energy-carbon-links-in-households-and-cities-a-new-paradigm.php.

Kenway, S. J., Binks, A., Bors, J., Pamminger, F., Lant, P., Head, B., ... Bader, H. P. (2014). Understanding and managing water-related energy use in Australian Households. *Water*, April, 184–188.

Kenway, S. J., Binks, A., Lane, J., Lant, P. A., Lam, K. L., & Simms, A. (2015). A systemic framework and analysis of urban water energy. *Environmental Modelling & Software, 73*, 272–285.

Kenway, S. J., Gregory, A., & McMahon, J. (2011a). Urban water mass balance analysis. *Journal of Industrial Ecology, 15*, 693–706.

Kenway, S. J., Lane, J., Binks, A., Lam, K. L., Simms, A., Reutter, B., ... Grinham, A. (2014). *Energy use, greenhouse gas emissions, and related costs associated with Seqwater infrastructure, and implications for the broader urban water system*. Brisbane. Retrieved from Project 3.3: Systemic Analysis of Energy Use and Greenhouse Gas Emissions through the Drinking Water Supply System in South East Qld.

Kenway, S. J., & Lant, P. A. (2012). The influence of water on urban energy use. (Chapter 5). In C. Howe & C. Mitchell (Eds.), *Water sensitive cities* (pp. 63–77). London: International Water Association.

Kenway, S. J., Lant, P. A., & Priestley, A. (2011b). Quantifying the links between water and energy in cities. *Journal of Water and Climate Change, 2*, 247–259.

Kenway, S. J., Lant, P. A., Priestley, A., & Daniels, P. (2011c). The connection between water and energy in cities - A review. *Water Science and Technology, 63*, 1983–1990.

Kenway, S. J., McMahon, J., Elmer, V., Conrad, S., & Rosenblum, J. (2013). Managing water-related energy in future cities - A research and policy roadmap. *Journal of Water and Climate Change, 4*, 161–175.

Kenway, S. J., Pamminger, F., & Lant, P. A. (2013). The case for urban metabolism. What is it and how can it help the water sector? *Water, February 2013*, 6–7.

Kenway, S. J., Priestley, A., Cook, S., Seo, S., Inman, M., & Gregory, A. (2008). *Energy use in the Provision and Consumption of Urban Water in Australia and New Zealand* (Report No. 9780643096165). Retrieved from CSIRO website: http://www.clw.csiro.au/publications/waterforahealthycountry/2008/wfhc-urban-water-energy.pdf.

Kenway, S. J., Scheidegger, R., Larsen, T. A., Lant, P. A., & Bader, H. P. (2013). Water-related energy in households: A model designed to understand the current state and simulate possible measures. *Energy and Buildings, 58*, 378–389.

Kenway, S. J., & Tjandraatmadja, G. (2009). Technological innovation in the provision of sustainable urban water services. In P. Newton, K. Hampson, & R. Drogemuller (Eds.), *Technology, design and process innovation in the built environment*. Abingdon, UK:Taylor Francis.

Kenway, S. J., Turner, G., Cook, S., & Baynes, T. (2008). *Water-energy futures for Melbourne: The effect of water strategies, water use and urban form*. Canberra: CSIRO.

Klein, G., Krebs, M., Hall, V., O'Brien, T., & Blevins, B. (2005). *California's water-energy relationship - Final staff report*. Sacramento, CA: California Energy Commission.

Lam, K. L., Kenway, S., & Lant, P. (2015). *Historical energy implications of urban water management options in South East Queensland, Australia*. 9th IWA Symposium on Systems Analysis and Integrated Assessment, Gold Coast, Australia.

Lane, J. L., de Haas, D. W., & Lant, P. A. (2015). The diverse environmental burden of city-scale urban water systems. *Water Research, 81*, 398–415.

Lesh, P. (2009). *Rate impacts and key design elements of gas and electric utility decoupling - A comprehensive review*. Retrieved from the Regulatory Assistance Project website: http://www.raponline.org/Pubs/Lesh-CompReviewDecouplingInfoElecandGas-30June09.pdf.

Lundie, S., Peters, G. M., & Beavis, P. C. (2004). Life cycle assessment for sustainable metropolitan water systems planning. *Environmental Science and Technology, 38*, 3465–3473.

Pamminger, F., & Kenway, S. J. (2008). Urban metabolism - improving the sustainability of urban water systems. *Water, 35*, 28–29.

Queensland Water Commission. (2010). *South east queensland water strategy*. Queensland Water Commission.

Racoviceanu, A. I., & Karney, B. W. (2010). Life-cycle perspective on residential water conservation strategies. *Journal of Infrastructure Systems, 16*, 40–49.

Racoviceanu, A. I., Karney, B. W., Kennedy, C. A., & Colombo, A. F. (2007). Life-cycle energy use and greenhouse gas emissions inventory for water treatment systems. *Journal of Infrastructure Systems, 13*, 261–270.

Roberts, P. (2012). *End use study, summer*. Melbourne: Yarra Valley Water.

Saliba, C., & Gan, K. (2006). *Energy density maps in water demand management*. Ozwater.

Satterthwaite, C. (2008). Cities' contributions to global warming: Notes on the alocation of greenhouse gas emissions. *Environment and Urbanization, 20*, 539–549.

Shrestha, E., Ahmad, S., Johnson, W., Shrestha, P., & Batista, J. R. (2011). Carbon footprint of water conveyance versus desalination as alternatives to expand water supply. *Desalination, 280*, 33–43.

Sterman, J. D. (1991). *A skeptic's guide to computer models*. Massachusetts Institute of Technology.

Stokes, J. R., & Horvath, A. (2006). Life cycle energy assessment of alternative water supply systems. *International Journal of Life Cycle Assessment, 11*, 335–343.

Stokes, J. R., & Horvath, A. (2009). Energy and air emission effects of water supply. *Environmental Science and Technology, 43*, 2680–2687.

Stokes, J. R., Horvath, A., & Sturm, R. (2013). Water loss control using pressure management: Life-cycle energy and air emission effects. *Environmental Science and Technology, 47*, 10771–10780.

Stolz, G.(2007, February 1). The Water Crisis - Its Desalinate or Die, *Courier Mail*, p. 1.

Sydney Water. (2010). *Annual Report 2010.* Retrieved from http://www.sydneywater.com.au/Publications/Reports/AnnualReport/2010/statutory_information/research_development.html

Sydney Water. (2012). Energy management & climate change website. Retrieved from http://www.sydneywater.com.au/SW/water-the-environment/what-we-re-doing/energy-management/index.htm.

Sydney Water. (2014). *Operating licence environment report*. Retrieved from Sydney Water website.

Victorian Water Industry Association. (2011). *Electricity issues in the Victorian Water Sector*. Melbourne: Victorian Water Industry Association.

WSAA. (2006). *Refilling the Glass*. Exploring the issues surrounding water recycling in Australia. Water Services Association of Australia.

WSAA. (2009). *WSAA Report Card 2008-2009. Performance of the Australian Urban Water Industry and projections for the future*. Melbourne: Water Services Association of Australia.

WSAA. (2012). *Cost of carbon abatement in the Australian water industry*. Retrieved from Water Services Association of Australia website: http://www.wsaa.asn.au/WSAAPublications/Documents/Occasional%20Paper%2028%20Cost%20carbon%20abatement%20in%20the%20urban%20water%20industry%20May%202012.pdf.

Energy for freshwater supply, use and disposal in the Netherlands: a case study of Dutch households

P.W. Gerbens-Leenes

Water Management Group, Twente Water Centre, University of Twente, Enschede, the Netherlands

ABSTRACT

This study presents energy requirements for Dutch household water: 10.2 GJ per capita per year, which includes 9.3 GJ (92%) for heating water, 0.6 GJ (6%) for water supply, and 0.2 GJ for wastewater treatment (2%). The top three energy consumers include shower water (58%), dishwasher water (9%) and washing machine water (8%). The Netherlands, a water-abundant country, expends far more energy to heat water for households than to supply municipal water, or to treat and dispose of wastewater. Policies to make water chains more sustainable should focus on use, rather than supply and disposal.

Introduction

The interrelationships between water and energy issues, often referred to as the 'water-energy nexus', have received much attention in the last decades. There are strong connections between the demand for and use of energy and water, with strong parallels between the growing water crises and conflicts over energy sources (Biswas & Tortajada, 2009; Gleick, 1994; Liu, Zhao, Gerbens-Leenes, & Guan, 2015; Mekonnen, Gerbens-Leenes, & Hoekstra, 2015; NETL, 2006). In 2005, a seminar on water and energy, organized within the overall framework of the World Water Week in Stockholm, addressed the interlinkages between energy and water (Tortajada & Varis, 2006). The seminar showed that water affects energy, and energy in turn is affected by water. Detailed studies on the energy requirements to supply water include those by Barrios, Siebel, van der Helm, Bosklopper, and Gijzen (2008) and Frijns, Mulder, and Roorda (2008), who gave an overview of energy requirements to supply freshwater and to dispose of wastewater in the Netherlands; Buckley, Friedrich, and Von Bloonitz (2011), who reviewed South African water sector life-cycle assessments; Plappally and Lienhard (2012), who gave an in-depth overview of energy requirements of water supply in the complete water chain, including energy for wastewater treatment, for the agricultural and municipal sectors in various countries; Sanders and Webber (2012), who evaluated the energy consumed for water use in the United States; and Williams and Simmons (2013), who gave an estimate of the energy requirements to supply water, including an estimate of the

Supplemental data for this article can be accessed at http://dx.doi.org/10.1080/07900627.2015.1127216

energy requirements for global water supply. In spite of the close interrelationships, however, the water profession as a whole has given inadequate attention to the energy sector (Tortajada & Varis, 2006). The provision of basic infrastructure services, such as water and sanitation, plays a vital part in the economic growth of a country and is essential for development (Tortajada, 2014). At present, the benefits of clean water and effective wastewater treatment are not available everywhere, as a consequence of the amount of both capital and energy needed to build and operate such systems (Plappally & Lienhard, 2012; Williams & Simmons, 2013). Estimates suggest that today the delivery of suitable water and the treatment of wastewater uses about 3% of the world's primary energy (Williams & Simmons, 2013). Further, end uses of water, such as water heating for the shower or dishwasher, also consume energy (Barrios et al., 2008; Frijns et al., 2008; Plappally & Lienhard, 2012).

Energy is required in the complete water chain. For example, energy is required to withdraw water (e.g. to pump groundwater), to purify water to the required quality level, to distribute water, to heat water in the use phase, to collect and treat wastewater, and to dispose of wastewater. Often, if water becomes scarcer or if the quality decreases, more energy is needed per unit of water supplied: water-scarce regions must often withdraw water from deep aquifers, and when water quality is low, energy-intensive treatment may be required. This means that local circumstances and choices made in every link of the water chain have a large impact on energy requirements of freshwater supply, use and disposal and therefore differ between locations. Case studies of specific countries and locations might contribute to better insight into the 'water-energy nexus' and show similarities and differences for energy requirements for freshwater supply, use and disposal. The Netherlands is a well-developed country with excellent water and energy infrastructure. Moreover, data on energy and water use are available. This makes it a good subject for a case study of the energy requirements for freshwater supply, use and disposal of its households.

Even though in some countries water is abundant, this does not necessarily mean that there is always enough freshwater of sufficient quality available for household supply, because sometimes droughts occur, or the water quality is insufficient. Although the Netherlands is located in the delta of large rivers, with roughly one-fifth of the country's surface consisting of water, it must consider freshwater availability for municipal water supply, because of the temporal variability of river runoff and changes in water quality. The occurrence of droughts or low-quality surface water sometimes constrains the ability to supply municipal water to households (Wolters-Noordhoff Atlas Productions, 2007). Moreover, demand for water is increasing. In the Netherlands between 1920 and 1990, the annual municipal freshwater use per capita increased from 17 to 70 m^3 per year (Wolters-Noordhoff Atlas Productions, 2007). Also, if total water use increases, more energy is needed to supply freshwater, to treat wastewater and to heat the water in the use phase.

Energy use has received much attention in the Netherlands. Especially in the last decade of the twentieth century, many studies have analyzed energy use by Dutch households (e.g. Falkena, Moll, Noorman, Kok, & Benders, 2003; Kramer & Moll, 1995; Kramer, Wiersma, Gatersleben, Noorman, & Biesiot, 1998). While those studies have given an in-depth overview of energy use in Dutch households, they have not specifically addressed the energy required for freshwater use. The water utilities in the Netherlands have also assessed energy use of the sector (Barrios et al., 2008; Frijns et al., 2008), providing detailed information on the energy requirements for municipal water for the Netherlands, including energy requirements for specific water treatment activities for municipal water. This article expands on these previous studies by estimating the energy embodied in 'water use' in addition to water extraction and treatment. The aim of this case study is to assess energy requirements related to Dutch household water supply, use and disposal for all components of the freshwater chain, including energy use in the household for heating, and to compare energy for water with total household energy use. Next, it compares the results with earlier studies to find general trends. By including energy for water use, the study provides a robust estimate of the total energy footprint associated with municipal water demand in the Netherlands.

Municipal water supply and wastewater treatment in the Netherlands

The Netherlands is a flat country in the delta of the Rhine, Meuse and Scheldt Rivers. It is water-abundant, as most of the renewable water resources it receives are never extracted: 77.8% flow to the sea, 17.6% evaporate and only 4.5% are consumed by humans (Wolters-Noordhoff Atlas Productions, 2007). Ten Dutch water utilities provide drinking water for municipal supply. They are located close to municipalities, and withdraw water from sources nearby. The purpose of this strategy is to ensure that costly, energy-intensive water transport systems are not required to meet municipal water demand (Rijksinstituut voor Volksgezondheid en Milieu, 2011). The water utilities select the highest-quality source for drinking water production (Smeets, Medema, & Van Dijk, 2009). The preferred source is groundwater, because it is microbiologically safe. In some parts of the country, however, groundwater is brackish and so surface water is used for drinking. Figure A1 (in the supplemental online material at http://dx.doi.org/10.1080/07900627.2015.1127216) shows the source water types for drinking water in the Netherlands (Smeets et al., 2009).

The purification processes used by each utility depend on the quality of the water source, the type of water extracted (ground or surface) and the drinking water standards required. Figure A1 shows that groundwater is used for drinking in most parts of the Netherlands. Groundwater in the Netherlands is high-quality water that is only minimally treated for substances such as ammonium, iron, manganese and oxygen. It is extracted from restricted areas owned by the water utilities. This decreases the risk of groundwater contamination (Smeets et al., 2009). For the purification of surface water, more treatment is needed, such as storage in a water reservoir, slow sand filtration, coagulation, UV disinfection or ozonation, and activated carbon filtration (Rijksinstituut voor Volksgezondheid en Milieu [RIVM], 2011). The quality of surface water in the Netherlands is determined by human activities upstream of the sources of extraction, e.g. wastewater discharges and industrial and agricultural activities. This upstream pollution can occur purposely through sanctioned industrial or agricultural activity or by accident. When water is accidentally contaminated, and the system is too heavily polluted to be used as a source of drinking water, water utilities use storage systems, including dunes. After persistent problems with polluted and brackish drinking water in the nineteenth century, the major towns in the west of the Netherlands eventually went to the groundwater reserves in the coastal dune belt for their water supply (De Vries, 2007). Figure A1 (in the supplemental online material) shows that in some areas surface water is used directly, and in other areas it is infiltrated in the dunes. Although most of the time the Rhine and Meuse provide sufficient water quantities, the water quality is not always sufficient. The surface water from the Meuse and Rhine is pretreated, transported to the dunes and infiltrated to improve its quality. The dunes form a natural filter for pathogenic micro-organisms, contribute to constant water quality and temperature, and have a large storage capacity to overcome pollution waves in the rivers (Smeets et al., 2009; RIVM, 2011).

In the Netherlands, the standards for surface water for drinking derive from a 1975 European Union directive (75/440/EEG) that gives critical concentrations for pesticides, polyaromatic hydrocarbons and chloride (RIVM, 2011). However, it is recognized that the directive does not consider all pollutants. Pollutants not included are hormones, medicines, nanoparticles (e.g. TiO_2), cosmetics, and food additives. The quality of Dutch surface water has improved since 1970 when considering the concentrations of dangerous pollutants from the directive of 1975. However, this has not resulted in lower requirements for purification, because of the introduction of other pollutants and more stringent water quality standards. This is due in part to better chemical analysis techniques to identify small concentrations of pollutants. The Dutch Ministry of Infrastructure and the Environment is making a new list of acceptable levels of pollutants in surface water for municipal use (RIVM, 2011).

At present, the Dutch policy standards for drinking water are included in the Drinking Water Law (Drinkwaterwet – Dutch Government, 2009). The protection of Dutch groundwater quality is regulated in the Water Directive, preventing groundwater pollution by setting standards for activities in areas where groundwater is withdrawn. The Dutch policy includes a complete ban on the use of chloride for purification (RIVM, 2011). In the Netherlands, per capita water use has decreased thanks to technological improvements, for example the introduction of more efficient toilets, washing machines and showers. These technological improvements, in combination with increased public awareness, were so effective that government campaigns to make the public more aware of their water use stopped (de Moel, Verberk, & van Dijk, 2006).

The Dutch Drinking Water Law (Dutch Government, 2009a) assigns the responsibility for wastewater treatment and disposal to the water boards. In 2006, 98.6% of total wastewater in the Netherlands was treated in a wastewater treatment plant (Directoraat-Generaal Water / Directoraat-Generaal Milieubeheer, 2008).

The freshwater chain

Figure 1 shows a freshwater chain that consists of six components: water withdrawal; water purification; water distribution; water use; collection of wastewater; and wastewater treatment and disposal. All of these components require energy. This article estimates the energy requirements of all six components for the Dutch water system. The study excludes the energy required to build infrastructure, such as pipes and pumps, as this is generally attributed to energy embodied in industry.

The total energy footprint of a freshwater chain depends on the choices made for each component. For example, surface water is more readily accessible than groundwater, and so less energy is required for its extraction. However, surface water generally contains more contaminants than groundwater, so more energy is required for treatment. The utility's choice of whether to extract surface or groundwater will be a function of the accessibility and purity of each; there is often a trade-off whereby more accessible water sources are also more contaminated. Another factor affecting the energy footprint of water is the desired quality. The desired quality of municipal water may differ between regions, which can affect the choice of purification method, and thus the energy costs. This is also the case with wastewater, as numerous options for treatment exist, each of which will bring water to a different level of purity while requiring a different amount of energy.

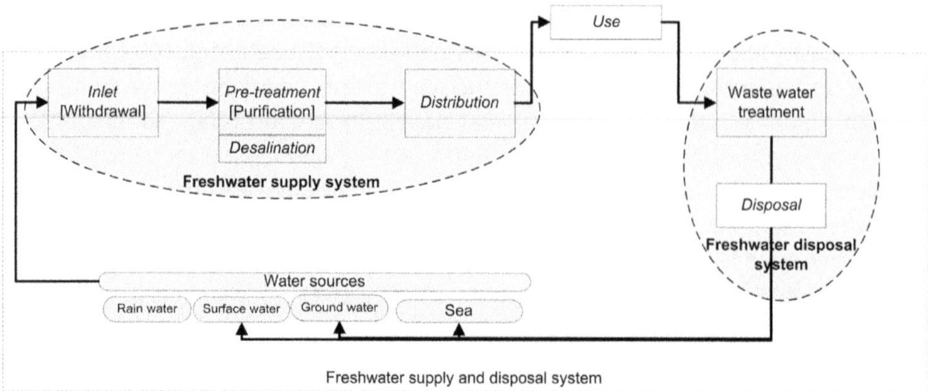

Figure 1. A typical freshwater chain, depicting its components.

Energy

The global energy system uses fossil fuels (coal, natural gas and conventional oil), renewables (e.g. biomass, solar energy, wind energy) and nuclear energy (uranium) for the supply of heat, electricity and transport fuels (IEA, 2015). The Netherlands has large natural gas stocks (Wolters-Noordhoff Atlas Productions, 2007), and not only exports but also uses large quantities of natural gas (IEA, 2015). Based on IEA statistics (2015), gas contributes about one-fifth of the total energy supply in the global energy mix. In the Netherlands the contribution is larger, at about one-third.

Energy is almost always required for the provision of water. Mechanical energy for pumping can be generated by humans and draught animals, renewable technologies, or diesel or electrical pumps. For example, either electrical or diesel energy is required to withdraw ground or surface water. Additionally, electricity is required for municipal water treatment. Examples of municipal water treatment that require electricity are biological activated carbon filtration, softening and ozonation (Barrios et al., 2008). Households also use energy for water. In the Netherlands, natural gas is one of the primary sources used for heating water. Last, when water is treated prior to disposal, energy is required.

The energy needed to produce useful fuels and electricity is termed 'energy required for energy' (ERE) and depends on the energy source and the efficiency of power generation. Values for electricity generated in the Netherlands are about 2.65 (Blok, 2006; Falkena et al., 2003). Different mixes for electricity have different ERE values. For example, the ERE of electricity from coal is 2.60 MJ/MJ, of electricity from natural gas 1.96 MJ/MJ (Meldrum, Nettles-Anderson, Heath, & Macknick, 2013), and for renewable energy, such as electricity from biomass, it is 2.5–5.0 MJ/MJ (Faaij, 2006), meaning that the latter form of electricity is almost twice as energy-intensive as electricity from coal or natural gas. When the energy requirements of water are assessed, these ERE values must be taken into account. This is because total energy use depends partly on the source of that energy. For example, when Dutch tap water is heated mainly by natural gas, and not by electricity, energy losses associated with the conversion of energy in a power plant are avoided.

Method

The study combined information on water use in Dutch households with data on energy for water (energy to supply water, to heat the water for a specific application, to dispose of water, and for wastewater treatment). In all the components of the freshwater chain, energy is needed for different processes. The total energy footprint of the freshwater chain (excluding the energy related to water infrastructure), EF_{water} (in MJ/m^3), consists of six components: $EF_{withdrawal}$, the energy footprint to withdraw the water from a specific source; $EF_{purification}$, the energy footprint to purify a unit of water; $EF_{distribution}$, the energy footprint to distribute the water to the required location; EF_{use}, the energy in the use phase, mainly for water heating; $EF_{collection}$, the energy footprint to collect water and bring it to a wastewater treatment plant; and $EF_{treat+disposal}$, the energy footprint to treat the wastewater to the required quality standard and then to dispose of it). This is shown in Equation (1):

$$EF_{water} = EF_{withdrawal} + EF_{purification} + EF_{distribution} + EF_{use} + EF_{collection} + EF_{treat+disposal} (MJ/m^3) \quad (1)$$

Information on water use in the Netherlands is available from TNS-NIPO (2008). Table 1 gives the average daily per capita household water use in the Netherlands for 1995–2007 for 12 use categories. For the assessment of the energy requirement in the use phase, this study used the data for 2007. Energy in the use phase is related to the heating of water. The study distinguished five categories of warm water use: washing machine; tap water (shower, bath, dishwashing by hand, personal washing, washing by hand); cooking; dishwashing machine; and coffee and tea. Total water use in a Dutch household in 2007 is expressed in m^3 per capita per year for five use categories in which the categories shower, bath, dishwashing by hand, personal washing and washing by hand from Table 1 are added into the category of warm tap water.

Table 2 gives the energy requirements of the processes that supply water for municipal services in the Netherlands from two Dutch sources, as well as energy for wastewater transport and treatment. The table shows that depending on the technology applied, energy requirements for water supply in the Netherlands might differ by a factor of three. When the ERE factor is included, municipal water supply ranges between 4 and 13 MJ/m^3. For sewage water transport and treatment, Frijns et al. (2008) give energy requirements of 1.05 and 3.35 MJ/m^3, respectively. The energy required to supply and dispose of water in the Netherlands lies between 8.4 and 17.4 MJ/m^3.

Table 1. Average water use (litres per capita per day) in the Netherlands, 1995–2007.

	1995	1998	2001	2004	2007
Toilet	42.0	40.2	39.3	35.8	37.1
Shower	38.3	39.7	42.0	43.7	49.8
Washing machine	25.5	23.2	22.8	18.0	15.5
Bath	9.0	6.7	3.7	2.8	2.5
Dishwashing by hand	4.9	3.8	3.6	3.9	3.8
Personal washing	4.2	5.1	5.2	5.1	5.3
Washing by hand	2.1	2.1	1.8	1.5	1.7
Cooking	2.0	1.7	1.6	1.8	1.7
Coffee/tea	1.5	1.1	1.0	1.0	1.2
Dishwasher	0.9	1.9	2.4	3.0	3.0
Drinking water	-	0.5	0.5	0.6	0.6
Other water use	6.7	6.1	6.7	6.4	5.3
Total	**137.1**	**131.9**	**130.7**	**123.8**	**127.5**

Source: TNS-NIPO (2008).

Table 2. Energy requirements of the processes to supply water for municipal services as well as energy for wastewater transport and treatment. All figures in kJ/m³ except where noted.

Processes	Barrios et al. (2006)	Frijns et al. (2008)
Raw water intake	40.4	2.2
Coagulation		86.4
Lake water reservoir		
Pumping from lake water	80.4	
Rapid sand filtration	49.1	43.2
Pumping to water treatment plant	100.4	
Ozonation	181.9	
Softening (energy for pumping)	574.9	72.0
Biological activated carbon filtration	0.0	28.8
Slow sand filtration	86.9	
Storage and pumping into distribution network	396.0	396.0
UV disinfection		223.2
Membrane filtration		3,600.0
Flotation		180.0
Aeration		288.0
Total electricity (kJ/m³)	1,510.0	4,919.8
Total energy (kJ/m³) including a factor for ERE (2.65)	4,001.5	13,037.5
Wastewater transport and treatment process		
Wastewater transport		1,050
Wastewater treatment		3,350
Total		4,400

In the use phase, households often heat the water, for example for showering or washing. To calculate the energy requirement related to freshwater use in a Dutch household, the study combined information on natural gas and electricity use for heating water from Kramer et al. (1998) with information on water use for bathing, showering, dishwashing and washing machines from TNS-NIPO (2008) for the year 1997. In 1997, an average Dutch household used 14.7 GJ to warm tap water (Kramer et al., 1998). The study assessed the household size in 1997 as the average of the household size in 1995 and 2000 of 2.35 and 2.30 persons, respectively (Centraal Bureau voor de Statistiek (CBS), Planbureau voor de Leefomgeving (PBL) en Wageningen Universiteit en Researchcentrum (Wageningen UR), 2015). It calculated energy use to warm tap water (in GJ per capita per year) by dividing household energy use for tap water by the household size in 1997 and by dividing energy use per capita in 1997 by the warm tap water use in 1998 from TNS-NIPO (2008). The study assumed that tap water use in 1998 was the same as in 1997. It also assumed that energy to warm tap water (in MJ/m³) did not change between 1997 and 2007.

To calculate the energy requirement for water for washing machines, the study derived data on water and electricity use from the Dutch Consumentenbond (2010), an organization that tests household machines. Based on the water and energy use per kg of clothes for three advised machines (Miele W5825, water use 13 L/kg, electricity use 0.13 kWh/kg; Whirlpool Denver 1600, water use 10 L/kg, electricity use 0.10 kWh/kg; Indesit IWC5145, water use 9 L/kg, electricity use 0.17 kWh/kg), the study calculated the electricity needed to heat a unit of water by dividing average energy use per kg of clothes (kWh/kg) by average water use per kg of clothes (m³/kg), converting kWh to MJ by applying a factor of 3.6 and converting litres to m³ by applying a factor of a thousand. The energy required to heat a unit of water (in MJ/m³) was calculated based on an ERE for electricity of 2.65.

For dishwashers, the study used the same approach and based the calculations on four brands: Ikea Lagan DW60, water use 14.83 L/turn, electricity use 1.064 kWh/turn; Bosch SRS4 5M02EU, water use 16.80 L/turn, electricity use 1.460 kWh/turn; Siemens SN65E001EU, water use 12.00 L/turn, electricity use 1.06 kWh/turn; and Whirlpool ADG 9527/2, water use 14.00 L/turn, electricity use 1.054 kWh/turn. The equation for energy for cooking was obtained from Gerbens-Leenes (2000):

$$\text{Energy for cooking} = \text{Vwg} \times \text{sww} \times \text{delta T} \times 1/\text{e} \times 10^{-6}(\text{MJ}) \quad (2)$$

where Vwg is the amount of water to be heated (g), sww is the specific heat of water (4.18 J/g·K) (Verkerk, 2008), delta T is the temperature difference between tap water and boiling water (assumed to be 85 °C), e is the efficiency of warming water using natural gas (0.5) and the factor of 10^{-6} is used to convert from J to MJ.

Results

Table 3 gives the energy requirements for warm tap water, washing and dishwashing machine water, cooking (based on natural gas use) and coffee/tea (based on electricity use), total energy use and total water use per capita per year in 2007. The energy used to heat water (MJ/m^3) for dishwashers is relatively large, while that for washing machines is relatively small, probably because washing machines use mostly cold water for rinsing. The table also shows that energy for water to prepare coffee or tea is the largest per unit of water, at roughly three times the energy requirement of warm tap water. This is caused by the difference in heating temperature and by the use of electricity rather than natural gas, which has a higher ERE. Energy to warm tap water dominates energy for water use; energy to prepare coffee and tea is the smallest. Most of the water is needed for tap water, followed by water for the washing machine. The amount of water used for coffee and tea is relatively small.

The results for the energy footprint of supply and disposal, 8.4–17.4 MJ/m^3, are based on the results for water supply of municipal services, including a factor of 2.65 for ERE. The study used the values of 4 and 13 MJ/m^3 for municipal water supply and of 4.4 MJ/m^3 for wastewater transport and treatment, based on data from Barrios et al. (2008) and Frijns et al. (2008) – see Table 2. When energy requirements for warm tap water, washing and dishwashing machine water, cooking (based on natural gas use) and coffee/tea (based on electricity

Table 3. Energy requirements for warm tap water, washing and dishwashing machine water, cooking (based on natural gas use) and coffee/tea (based on electricity use), plus total energy use and total water use per capita per year, for 2007.

	Energy for heating (MJ/m^3)	Energy for supply and disposal (MJ/m^3)	Total energy (MJ/m^3)	Total energy use (MJ per person per year)	Total water use (m^3 per person per year)
Washing machine	124	8.4–17.4	132–141	747–798	5.66
Warm tap water	302	8.4–17.4	310–319	7144–7352	23.03
Cooking (natural gas)	711	8.4–17.4	719–728	446–452	0.62
Dishwashing machine	769	8.4–17.4	777–786	851–861	1.10
Coffee/tea (electricity)	942	8.4–17.4	950–959	416–420	0.44

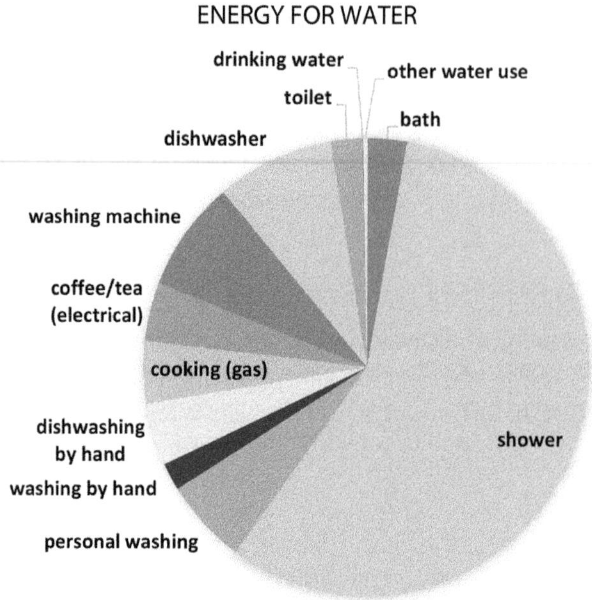

Figure 2. Energy for water in Dutch households.

use) are combined with water use, the energy requirement for municipal water supply and use in the Netherlands lies between 9.6 and 10.2 GJ per capita per year. When the higher value for water supply from Frijns et al. (2008) is taken, 9.3 GJ (92%) is energy used by households to heat water, 0.6 GJ (6%) for water supply, and 0.2 GJ for water treatment (2%). The energy requirement for heating water is dominated by the shower (on average 5.5 GJ per capita per year), the dishwasher (0.8 GJ per capita per year) and the washing machine (0.7 GJ per capita per year). Energy for heating the rest of the water in a Dutch household (washing, dishwashing, cooking, and preparing coffee and tea) is 2.3 GJ per capita per year. Figure 2 shows the relative energy use per household function. The top three energy requirements are for water for the shower (58%), water for the dishwasher (9%) and water for the washing machine (8%).

Discussion

The data for Dutch municipal water supply and disposal used in this study are derived from the detailed studies of Barrios et al. (2008) and Frijns et al. (2008). Energy requirements for water supply, including ERE values, vary between 4 and 13 MJ/m^3; wastewater transport and disposal requires 4.4 MJ/m^3 (including ERE). In the Netherlands, water leakages in the supply system are small, in general less than 3% (Smeets et al., 2009). In countries where leakage rates are high, for example in Bulgaria, where the leakage rate is 50% (European Environment Agency, 2003), energy to supply a unit of water to a household (MJ/m^3) doubles. When energy for water is compared for different countries, losses also need to be taken into account. A recent study of the energy footprints of water production, treatment, end use, reclamation and disposal (Plappally & Lienhard, 2012) gave values between 0.01 kWh/m^3 (Australia) and 1.5 kWh/m^3 (Spain) for conventional water treatment. This is 0.04–5.2 MJ/m^3 without the ERE value and 0.1–14.3 MJ/m^3 including the ERE value. The values for the Netherlands fall within this range. A recent study from South Africa on the life cycle of the water sector (Buckley et al., 2011) gives similar values. For example, for the primary treatment of wastewater, both Frijns et al. and Buckley et al. give energy requirements of 0.11 kWh/m^3.

This study finds that in the Netherlands energy use related to water use in households is dominated by the heating of water. Energy is used to warm water for clothes washers, dishwashers, taps, baths and showers. Energy requirements per m^3 of water depend on the source of the energy for heating, in which natural gas is more efficient than electricity, and on the final temperature of the water. For cooking, water needs to be heated to its boiling temperature (100 °C), requiring more energy (MJ/m^3) than water that needs to be heated for the shower. This study finds energy requirements ranging from 124 MJ/m^3 to heat the water for the washing machine to 942 MJ/m^3 to heat the water for coffee and tea preparation. Estimates for California, in the US, show that the warming of one cubic metre of water ranges between 5.4 kWh (19.4 MJ/m^3, including ERE for gas for the shower) and 25.9 kWh (247 MJ/m^3, including ERE for electricity) for the dishwasher (Plappally & Lienhard, 2012), which is much more than the energy needed to purify water, even when using desalination technologies. However, there is a large difference between the values for heating shower water in the Netherlands and in California. The specific heat of water is 4.18×10^3 J kg^{-1} K^{-1} (Verkerk, 2008). This means that in the Netherlands, where heating of the shower water requires 302 MJ/m^3, when no energy losses occur, the temperature difference between water before heating and after heating is about 72 K, whereas in California the temperature difference is only 5 K. Assuming that shower water does not have temperatures far above the body temperature of 38 °C, possibly the values for the Netherlands are overestimated or the heating of tap water is energy-inefficient. For cooking, the efficiency to heat water using natural gas is about 50% (Gerbens-Leenes, 2000). The results for the heating of tap water suggest that the efficiency is smaller than for cooking. On the other hand, the temperature increase for California is small. Because the impact of the use of warm water for the shower on the total energy use in a household is so large, better estimates of the energy to heat shower water are needed.

This study finds that 92% of the energy use for water in a Dutch household goes to heat the water. This is larger than the results of other studies, probably because this study also took the ERE values into account. Frijns et al. (2008) estimated that the energy use in the water use stage of the Dutch municipal water supply chain is four times the total for the other steps of the chain. This estimate of Frijns et al. is in line with two other studies that give similar results for the US. Sanders and Webber (2012) show that 75% of energy for water in US households is used to heat the water. Plappally and Lienhard (2012) show that in California 72% of the energy is used in households to heat the water. This means that in order to decrease the energy footprint of the water supply chain, the focus should be on the consumption stage, where human behaviour, especially related to showering and bathing, is the most important factor. The energy required for specific technologies in different steps of the water supply chain has received much attention, not only from scientists, but also from the water sector itself. Where technologies of municipal water suppliers in some countries are well documented, as was shown for the Netherlands (Barrios et al., 2008; Frijns et al., 2008), information for many other countries for different types of water sources is lacking or not publicly available. For example, the FAO's database, Aquastat, does not provide information on wastewater treatment for many EU countries for 1993–2002 (FAO, 2015). This means that the results of this case study cannot easily be compared to other countries or to other sectors.

Where utilities in the water sector providing municipal water have a good overview on how much water is supplied to customers and how it is treated, there is limited information on agricultural and industrial water supply. For example, there are large differences in energy footprints for groundwater pumping that depend on groundwater tables and on pumping efficiencies. The pumping efficiency of 25% reported for South Asia (Shah, Scott, Kishore, & Sharma, 2003) is relatively low compared to efficiencies of 70% published by the Hydraulic Institute, Europump, and the US Department of Energy's Office of Industrial Technologies (2000).

It is stressed that the results presented here are based on rough estimates of the freshwater supply chain and should be interpreted not at face value but within the context of all assumptions made. For water, the study used information from a limited number of sources. For energy, it combined information from the literature. The results, therefore, give an indication of energy footprints of water, show differences among types of water used and treated, and estimate per capita energy footprints of water in Dutch households. Although the study has used rough estimates, it gives a first impression of the relative energy footprint of water supply in comparison with total energy supply in a household and can be used to estimate future energy consequences from changes in the water system. When more comprehensive data-sets become available, it will be possible to make more accurate calculations. In order to contribute to the debate on the 'water-energy nexus', this study applies a systems approach giving the factors that determine the energy footprint of the water supply chain, from water withdrawal to disposal. It shows that, basically, there are six steps in the water supply chain that determine the energy footprint of water: water withdrawal, purification, distribution, use, wastewater collection, and wastewater treatment and disposal. In each of these steps, policy choices need to be made that will not only influence the energy footprint per step, but also cause trade-offs in other links. For example, the intake of surface water requires less energy than the intake of groundwater, but often surface water needs more energy for purification. The study shows that human behaviour in the water use stage dominates energy footprints in the whole water supply chain. If less water is used for energy-intensive processes, like showering, then significant gains can be made. In the European Union, labelling requirements show consumers the water and energy efficiency of household machines, such as washing machines (Milieucentraal, 2015). This could contribute to lower water and energy use.

Conclusions

The Netherlands is a country with abundant water that uses groundwater and surface water for freshwater supply. Although most of the year large rivers provide sufficient water, sometimes droughts occur, necessitating storage, and surface water quality is not always sufficient. Water quality is improved by infiltration in dunes; chlorine is not used at all. The Dutch policy in supplying municipal water is one of regional self-sufficiency, so that utilities do not need to transport water over large distances or elevate it to higher levels. The result of this policy is that water utilities obtain water from areas close to municipal demand centres. In the withdrawal, purification and distribution components of the water chain, relatively little energy is required, about 8–17 MJ per cubic metre of water produced. This figure includes wastewater disposal and treatment. In the Netherlands, in 1995–2007, average per capita water use decreased from 137 to 127 litres per day. This reduction was mainly caused by more efficient toilets and washing machines, in combination with public-awareness campaigns that emphasized the importance of water conservation. On the other hand, water use for showers increased from 38 to 50 litres per capita per day. This trend is important from an energy perspective, because showers represent 58% of total energy requirements for Dutch household water use.

In the Netherlands, the energy requirement of municipal water lies between 9.6 and 10.2 GJ per capita per year, dominated by direct household energy use for heating water. The top three energy requirements are energy for shower water (58%), energy for washing machine water (9%) and energy for dishwasher water (8%). When one compares energy for water to total energy requirements per capita per year of 112 GJ (Moll et al., 2005), the contribution of energy for water is about 9% of the total direct and indirect energy requirements of households in the Netherlands. This suggests that there are potential system gains from reducing the energy required to meet water demand.

To reduce the energy required to meet water demand, one should consider not only the energy required for water production, including energy losses due to leakages, but also the energy associated with how water is used. In the Netherlands, water use represents a significantly larger portion of energy consumption than water supply and wastewater treatment. Ninety-two per cent of energy for water is used to heat water. In the US it has been shown that 72–75% of energy for water in US households is needed to heat the water. While these case studies offer insights into energy for water in the Netherlands and the US, similar case studies should be performed in other countries to allow comparison. Case studies from water-scarce countries that require significantly more energy for water extraction would be particularly relevant for comparison, as they would demonstrate whether the relationship between energy for water supply and water use differs between water-scarce and water-abundant countries. The conclusions suggest that the energy associated with water use is much larger than the energy for water supply and wastewater treatment and disposal. Given this, any initiatives aimed to reduce energy consumption related to water should focus on the end use of water in addition to how water is supplied.

Acknowledgements

I would like to thank the editorial staff of the IJWRD and the external reviewers for the excellent comments offered for my journal submission. My special thanks go to Christopher Napoli for all his suggestions to improve the article. All these thoughtful comments have contributed to the improvement of earlier versions of the article.

Disclosure statement

No potential conflict of interest was reported by the author.

References

Barrios, R., Siebel, M., van der Helm, A., Bosklopper, K., & Gijzen, H. (2008). Environmental and financial life cycle impact assessment of drinking water production at Waternet. *Journal of Cleaner Production, 16*, 471–476.

Biswas, A. K., & Tortajada, C. (2009). Changing global water management landscape. In A. K. Biswas, C. Tortajada, & R. Izquierdo (Eds.), *Water management in 2020 and Beyond* (pp. 1–34). Berlin, Germany: Springer.

Blok, K. (2006). *Introduction to energy analysis*. Amsterdam, the Netherlands: Techne Press.

Buckley, C., Friedrich, E., & Von Bloonitz, H. (2011). Life-cycle assessments in the South African water sector: A review and future challenges. *Water SA, 37*, 719–726.

Centraal Bureau voor de Statistiek (CBS), Planbureau voor de Leefomgeving (PBL) en Wageningen Universiteit en Researchcentrum (Wageningen UR). (2015). *Compendium voor de Leefomgeving.* Retrieved from http://www.compendiumvoordeleefomgeving.nl/overhetclo/

Consumentenbond. (2010). Vergelijk alle wasmachines. Retrieved November 2, 2010, from http://www.consumentenbond.nl/test/woning-huishouden/huishouden/wasmachines/

De Moel, P. J., Verberk, J. Q. J. C., & van Dijk, J. C. (2006). *Drinking water. Principles and practices*. Singapore: World Scientific Publishing.

De Vries, J.J. (2007). Groundwater. In Th.E. Wong, D.A.J.Batjes & De Jager, J. (Eds.), *Geology of the Netherlands* (pp. 295-315). Amsterdam, the Netherlands: Royal Netherlands Academy of Arts and Sciences.

Directoraat-Generaal Water / Directoraat-Generaal Milieubeheer. (2008). *Inzameling, transport en behandeling van afvalwater in Nederland; situatie per 31 december 2006.* The Hague, The Netherlands: VROM.

Dutch Government. (2009). *Drinkwaterwet*. Retrieved from http://wetten.overheid.nl/BWBR0026338/.

European Environment Agency. (2003) Indicator Fact Sheet (WQ06) Water use efficiencies (in cities: leakage. Retrieved from http://www.eea.europa.eu/data-and-maps/indicators/water-use-efficiency-in-cities-leakage#toc-2

Faaij, A. (2006). Modern biomass conversion technologies. *Mitigation and Adaptation Strategies for Global Change, 11*, 335–367.

Falkena, H. J., Moll, H. C., Noorman, K. J., Kok, R., & Benders, R. M. J. (2003). *Household metabolism in Groningen*, Dutch national report – Groningen (deliverable no. 8). IVEM- research report 109. Groningen, the Netherlands: IVEM, University of Groningen.

FAO. (2015). *Aquastat.* Retrieved February 2015, from http://www.fao.org/nr/water/aquastat/data/query/index.html;jsessionid=9D346F54D8C5566507B492575088E5AE

Frijns, J., Mulder, M., & Roorda, J. (2008). *Op weg naar een klimaatneutrale waterketen, STOWA 2008–17.* The Netherlands: STOWA.

Gerbens-Leenes, P. W. (2000). *Groen Kookboek. Milieubewust koken met een laag energie- en landgebruik.* IVEM-onderzoeksrapport 103. Groningen, the Netherlands: IVEM, University of Groningen.

Gleick, P. H. (1994). Water and Energy. *Annual Review of Energy and the Environment, 19*, 267–299.

Hydraulic Institute, Europump, US Department of Energy's Office of Industrial Technologies (OIT). (2000). Pump life cycle costs: a guide to LCC analysis for pumping systems. Executive summary.

International Energy Agency. (2015). *Key world energy statistics*. Paris: International Energy Agency. Retrieved from http://www.iea.org/.

Kramer, K. J., & Moll, H. C. (1995) *Energie voedt. Nadere analyses van het indirecte energieverbruik van voeding. IVEM-onderzoeksrapport no. 77*. Groningen, The Netherlands: IVEM, Rijksuniversiteit Groningen.

Kramer, K. J., Wiersma, G., Gatersleben, B. C. M., Noorman, K. J. & Biesiot, W. (1998). *Huishoudelijk energiegebruik gemeten: aard, omvang en veranderingspotentieel. IVEM-onderzoeksrapport no. 93.* Groningen, The Netherlands, IVEM, Rijksuniversiteit Groningen.

Liu, J., Zhao, D., Gerbens-Leenes, P. W., & Guan, D. (2015). China's rising hydropower demand challenges water sector. *Scientific Reports, 5*, 11446. doi:10.1038/srep11446.

Mekonnen, M. M., Gerbens-Leenes, P. W., & Hoekstra, A. Y. (2015). The consumptive water footprint of electricity and heat: a global assessment. *Environmental Science: Water Research & Technology, 1*, 285–297.

Meldrum, S., Nettles-Anderson, S., Heath, G., & Macknick, J. (2013). Life cycle water use for electricity generation: a review and harmonization of literature estimates. *Environmental Research Letters, 8*, 015031.

Milieucentraal. (2015). Energielabel. Retrieved from www.energielabel.nl

Moll, H. C., Noorman, K. J., Kok, R., Engström, R., Throne-Holst, H., & Clark, C. (2005). Pursuing more sustainable consumption by analyzing household metabolism in European countries and cities. *Journal of industrial ecology, 9*, 259–275.

NETL. (2006) *Water & energy. Addressing the critical link between the nation's water resources and reliable and secure energy*. Albany: US Department of Energy, Office of Fossil Energy, National Energy Technology Laboratory.

Plappally, A. K., & Lienhard, J. H. (2012). Energy requirements for water production, treatment, end use, reclamation, and disposal. *Renewable and Sustainable Energy Reviews, 16*, 4818–4848.

Rijksinstituut voor Volksgezondheid en Milieu (RIVM). (2011) Toekomstverkenning drinkwatervoorziening in Nederland. RIVM rapport 609716001/2011. RIVM, Bilthoven, the Netherlands.

Sanders, K. T., & Webber, M. E. (2012). Evaluating the energy consumed for water use in the United States. *Environmental Research Letters, 7*,034034 (11pp)

Shah, T., Scott, C., Kishore, A., & Sharma, A. (2003). Energy-irrigation nexus in South Asia. Improving groundwater conservation and power sector viability. Research report 70. Sri Lanka: International Water Management Institute (IWMI).

Smeets, P. W. M. H., Medema, G. J., & Van Dijk, J. C. (2009). The Dutch secret: how to provide safe drinking water without chlorine in the Netherlands. *Drinking Water Engineering and Science, 2*, 1–14.

TNS-NIPO. (2008) *Watergebruik thuis*. Amsterdam, the Netherlands: TNS-NIPO.

Tortajada, C. (2014). Water infrastructure as an essential element for human development. *International Journal of Water Resources Development, 30*(1), 8–19. doi:10.1080/07900627.2014.888636.

Tortajada, C. & Varis, O. (2006). Seminar on water and energy, Stockholm, Sweden, August 21, 2005. *Water International, 31*(1), 133–134.

Verkerk, G. (2008). *Binas, (English edition) Handbook for the natural sciences and mathematics*. Groningen, the Netherlands: Wolters-Noordhof bv.

Williams E. D., & Simmons, J. E. (2013). *Water in the energy industry, An introduction*, BP International Ltd. Retrieved from www.bp.com/energysustainabilitychallenge

Wolters-Noordhoff Atlas Productions. (2007). *De Bosatlas van Nederland*. Groningen, the Netherlands: Wolters-Noordhof Atlas Productions.

Evaluating the economic viability of solar-powered desalination: Saudi Arabia as a case study

Christopher Napoli and Bertrand Rioux

King Abdullah Petroleum Studies and Research Center, Riyadh, Saudi Arabia

ABSTRACT

This article constructs a cost calculator to estimate the economic competitiveness of solar-powered desalination in Saudi Arabia. Solar desalination is defined as a plant that obtains solar energy from a closed system. This is done to focus the investigation on desalination technologies, rather than the efficacy of replacing conventional energy sources with renewables in an integrated electricity grid. The results suggest that current options for solar-powered desalination are not cost-competitive compared to incumbent technologies in Saudi Arabia. The article offers insight into where costs must decrease before solar technologies are economically competitive in the country.

As societies become more populous and wealthy, many increasingly place strains on their natural resources. This phenomenon is evident in Saudi Arabia, a country endowed with few natural resources that has experienced tremendous population and economic growth over the past half-century. Since 1960, the population of Saudi Arabia has risen from roughly 4 million to 29 million (World Bank, 2014a), while GDP per capita grew from USD 780 to USD 25,850 between 1968 and 2013 (World Bank, 2014b).

The growth in population and wealth has led to substantial increases in many types of consumption. For example, historically agriculture in Saudi Arabia was limited to date farming, small-scale herding and vegetable production. In the 1970s, however, the government created an intensive agriculture subsidy regime. The goals of the subsidies were to facilitate the modernization of farming techniques to meet increasing domestic demand and to improve the standard of living and employment opportunities in rural areas (Ouda, 2013). This resulted in an 18-fold increase in cereal production, an 8-fold increase in alfalfa production, and a 7-fold increase in meat production between 1973 and 2000 (Elhadj, 2004).

Similarly, and in part because of the growth in agricultural production, Saudi Arabia's water use has increased alongside population and wealth. In 1975, Saudi Arabia used roughly 1.75 km^3 of water. By 2006, that figure had risen to 23.67 km^3 (Aquastat, 2014). It is important to note that roughly 70% of the water withdrawn in Saudi Arabia comes from non-renewable fossil aquifers located deep within the earth. At current consumption trends, Saudi Arabia's renewable water resources per capita are depleting by 2% per year (Giansiracusa, 2010).

To combat the rising demand for water and depleting reserves, Saudi Arabia uses advanced treatment technologies to convert saline water into freshwater. In 2012, roughly 1.5 km^3 of desalinated water was distributed across municipal water networks for industrial, commercial and residential use in Saudi Arabia, making the country the largest producer of desalinated water in the world (Ministry of Water & Electricity, 2013). The development of large-scale desalination is aided by the country's expansive energy resources, which are sold domestically to utilities at low administered prices (Abderrahman, 2010).

While desalination has somewhat reduced the strain on non-renewable aquifers, it comes at a significant financial and environmental cost. Financially, Saudi Arabia used roughly 28 million barrels of oil equivalent (boe) to desalinate all water distributed to municipal and industrial water networks in 2011 (Matar, Murphy, Pierru, & Rioux, 2014b). As oil and natural gas are sold domestically at administered prices much lower than their international market values, this domestic use comes at a large opportunity cost – i.e. the difference between the domestic sale price and the price that could be obtained by selling the product on the international market at higher world prices.

This point is of particular importance given current oil consumption trends in Saudi Arabia. In the last 40 years, oil consumption in Saudi Arabia has risen by an average of 5.7% annually, resulting in a ninefold increase. This has made Saudi Arabia the sixth-largest consumer of oil globally, trailing only five (much larger) economies: the US, China, India, Japan and Russia (Gately, Al-Yousef, & Al-Sheikh, 2012). These trends have led some to suggest that, despite Saudi Arabia's currently being the world's largest crude oil exporter, it could become a net importer of oil by 2030 (Daya & El Baltaji, 2012). Given that oil represents 86% of the country's annual revenue, managing the long-term sustainability of the resource is of critical importance. Likewise, from an environmental perspective, using fossil fuels as energy sources for desalination, particularly energy-intensive thermal technologies like multistage flash (MSF), have increased the carbon footprint of the country.

Given the financial and environmental concerns, policy makers have begun examining how renewable energy sources, and in particular solar technologies, might be used to power desalination plants. A number of studies have tried to address this question. For example, Qiblawey and Banat (2008) described how direct solar desalination could be used for small-scale output in rural areas. Similarly, Gude, Nirmalakhandan, Deng, and Maganti (2012) offered a detailed analysis of the feasibility of using solar collectors augmented by thermal energy storage for low-temperature desalination. In 2012, an analysis by the World Bank (2012) explored the extent to which concentrated solar power (CSP), which can be stored and used during times of low sunlight, could be used to power large-scale desalination plants around the clock. The study concluded that CSP had tremendous potential to provide energy security and reduce greenhouse gas emissions, particularly in the Middle East and North Africa.

While there has been research on the engineering of solar-powered desalination, less attention has been paid to the economic costs and benefits. This article introduces the KAPSARC Cost Calculator for estimating the efficacy of using solar power as an energy source for desalination. The derived costs are then compared to the costs of desalination using incumbent technologies. The conclusions estimate whether current solar desalination options are cost-competitive with incumbent technologies, and which technological improvements may improve the attractiveness of solar-powered desalination.

It should be noted that the costs and operational parameters for water desalination can vary significantly, depending on plant configuration as well as the costs of labour and energy. The costs used in this article are taken from recent desalination infrastructure projects in Saudi Arabia. Given that costs may change based on project size and local conditions, the primary objective of this article is not to calculate the economic attractiveness of specific projects so much as to illustrate a framework for comparing the costs of different technologies. The calculator allows users to insert their own cost and other economic data. The benefits of this approach are twofold. First, as time changes the costs of technologies, these changes can be placed into the calculator to find new costs of each desalination method. Second, by offering a harmonized approach for calculating the cost of different desalination technologies, the framework can be used by other water-scarce regions, allowing international comparison of the costs of solar-powered and conventional desalination technologies.

Desalination in Saudi Arabia

Saudi Arabia has a long history of desalination. In 1938, the world's first large-scale desalination plant was established in Jeddah, Western Region. However, it was not until the 1970s, when windfall profits from oil exports made the process financially viable on a large scale, that the country truly adopted desalination as a method for meeting national water demand. It should be noted that desalination was strategically selected as the option for satisfying current and future domestic water supply requirements over alternatives such as large water transfer projects and heavy dependence on deep groundwater extraction (Al-Ibrahim, 2013). The choice of desalination was due to the country's expanding oil production (which negated the high energy costs), the geographical limitations of large water transfers, and a fear that over-reliance on deep groundwater withdrawals was not sustainable in the long term. The strategy to employ desalination technologies was executed with the 1974 Royal Decree No. R/49, which led to the creation of the Saline Water Conversion Corporation as an independent public body. The corporation was given the resources and mandate to increase the presence of desalination plants throughout the country. This mandate led to a 100-fold increase in the production of desalinated water by the 1990s. By 2010, there were 1595 desalination plants operating in Saudi Arabia (including both private and state-run plants), and roughly 50% of municipal water demand was being met with desalination technologies. Today, 57% of the output capacity for desalination in Saudi Arabia comes from thermal technologies, namely MSF (46.8%) and multi-effect distillation (MED, 10.3%), while roughly 40% comes from reverse osmosis (RO), a membrane process. The remaining 3% comes from advanced desalination technologies such as nanofiltration and membrane bioreactors (own calculations, data from Desal Database, 2014).

The reasons for the dominance of thermal technology in Saudi Arabia are threefold. First, because thermal technologies use the simple process of distillation to separate freshwater from salt and other impurities, it has historically been a more reliable method for large-scale desalination (Bernat, Gibert, Guiu, Tobella, & Campos, 2010), particularly in the highly saline seawater of the Gulf region (where the salt content is roughly 45,000 ppm, compared to the global average of 35,000 ppm). Second, thermal desalination can use waste heat from

Table 1. Baseline and solar desalination scenarios.

	Desalination type	Description	Pros	Cons
1	RO grid-powered (baseline)	RO powered by an electric grid	Low overall energy consumption	High operations and maintenance costs, membrane fowling, reliability issues under high salt concentrations and other impurities
2	RO solar PV with electric storage	RO powered by a solar PV plant with electric power storage	No fuel consumption	High capital cost (renewable power and electric storage), complex system integration
3	RO solar PV with water storage	RO powered by a solar PV plant with water storage	No fuel consumption	High capital cost (renewable power and spare desalination capacity), variable power supply, complex system integration
4	RO solar CSP with water storage	RO powered by a CSP plant with thermal and water storage	No fuel consumption	High capital cost (renewable power and spare desalination capacity), variable power supply, complex system integration
5	MED cogeneration (baseline)	MED plant with a combined-cycle gas turbine cogeneration power plant	Lower operations and maintenance costs, less sensitive to salt concentrations	High energy consumption and capital cost
6	MSF cogeneration (baseline)	MSF plant with a combined-cycle gas turbine cogeneration power plant	Lower operations and maintenance costs, less sensitive to salt concentrations	High energy consumption and capital cost
7	MED solar	MED plant powered by an inexpensive solar heat collector	Simple direct solar power	High heat demand and high overall systems cost

Note: RO = reverse osmosis; PV = photovoltaic; CSP = concentrated solar power; MED = multi-effect distillation; MSF = multistage flash.

cogeneration power plants, thus allowing water and energy needs to be met simultaneously from one utility.

Last, and most importantly, when energy prices are very low, thermal technologies, which are more energy-intensive but less capital- and labour-intensive, produce water more cheaply than membrane technologies like RO. Given the salinity of the Gulf region, seawater must often pass through two or three membranes before its salt content is completely removed, which increases the overall capital costs and non-fuel operating costs of membrane desalination plants (Saif, 2012).

As stated, while desalination has helped Saudi Arabia combat water scarcity, this has come at a significant financial and environmental cost, particularly given the dominance of energy-intensive thermal technologies. As a result, increased emphasis is now being placed on how renewable resources might be used to harmonize Saudi Arabia's desire to meet increasing water demand with financial and environmental objectives. For example, a solar desalination plant is being constructed in the town of Al-Khafji, in the north-east of Saudi Arabia. It is designed to provide 60,000 m^3 of desalinated water per day using RO with a solar photovoltaic (PV) plant capable of supplying the power for the desalination process (Abengoa, 2015).

Cost calculator description

A calculator has been constructed to estimate the unit cost of producing a cubic metre of potable water using conventional and solar desalination technologies. The calculator takes into account the capital costs, non-energy variable costs and energy costs of each technology. All costs are discounted over the technology's anticipated lifetime and converted to US dollars per cubic metre of water.

This article defines *solar desalination* as a desalination plant that obtains solar energy from a closed system. Thus, the article does not assess the efficacy of using solar power to supplement power obtained from an electricity grid. This strict definition was chosen in order to isolate the discussion to the costs and benefits of solar power, which requires excess investment in capacity and storage to ensure that water is provided even when the sun is not shining. Supplementing power from the grid with solar power does not isolate this discussion, as it is effectively akin to having a power station replace conventional fuels with solar options.

The costs of four different solar-powered desalination techniques are compared with three baseline scenarios: RO grid-powered, MSF cogeneration and MED cogeneration. The seven scenarios are described in Table 1.

Solar and conventional reverse osmosis scenarios

In Scenario 1, a baseline scenario, the cost of conventional RO desalination is calculated for an open system powered by an electric grid. RO has a low capital cost, but it requires relatively high materials costs to maintain the membranes used to separate salt and other impurities from the feed water. Pressure is applied across the membrane with electrical pumps, requiring a stable source of electrical energy.

Energy costs are defined by the price paid for electricity from the local grid provider. This scenario represents the standard way RO operates in Saudi Arabia. It should be noted that this scenario could also be used to represent RO powered by an off-grid conventional power supply, such as the diesel generators typically used in remote areas.

In general, solar desalination using RO technology can be achieved in two ways. Both involve investment in sufficient renewable power capacity to replace the consumption of conventional fuels. The first approach involves investment in *electricity storage*, so the desalination unit can operate at full capacity when the solar power output is below the peak (Scenario 2, RO solar PV with electric storage). A second approach is investing in spare desalination capacity and using *water storage* to capture the excess water produced during the day. The article assesses two different solar options involving water storage: Scenario 3 (RO solar PV with water storage) and Scenario 4 (RO solar CSP with water storage). Each option has costs and benefits: PV provides a lower capital cost to supply solar power, while the thermal storage provided by CSP increases the overall capacity factor of the power element.

Solar and conventional thermal desalination scenarios

The baselines for conventional thermal desalination are represented by Scenario 5 (MED cogeneration) and Scenario 6 (MSF cogeneration). These technologies heat the feed water, producing steam. The steam is then condensed back into a pure source of water, recovering some of the heat back into the feed water. These thermal distillation technologies typically use a conventional boiler, but nuclear and other renewable heat sources can also be used. Compared to MED, MSF has higher capital and operating costs. Both require significantly more energy than RO to boil the feed water, but at lower operating costs. As mentioned earlier, when primary energy prices are low enough, the cost of thermal desalination can drop well below that of modern RO.

In each scenario, a power and desalination plant are combined to provide both electricity and water. This coupling, known as cogeneration, has the added benefit of diversifying a utility's revenue stream. In a market such as Saudi Arabia, where revenues from the sale of electricity are greater than from providing freshwater, cogeneration provides a commercial advantage over stand-alone desalination plants. Although MSF is typically more expensive and energy-intensive than MED, in cogeneration an MSF plant will provide a higher power-to-water ratio, increasing overall electricity production and related revenues. The equivalent electrical energy of desalination, a measure of the useful electrical energy extracted from the steam turbine for desalination, is used to calculate the fuel cost of cogenerated water.

Last, Scenario 7 (MED solar) represents solar-powered thermal desalination. In this scenario, solar heat collectors are used to power an MED plant. A review of seawater desalination by Kalogirou (2005) references several pilot studies for solar-powered distillation using flat-plate, parabolic, and solar pond heat collectors. A unit cost calculator was not introduced for this scenario. A recent techno-economic analysis estimated that MED powered by evacuated-tube solar thermal collectors could cost as little as RMB 26/m^3, or close to USD 4/m^3 (Liu, Chen, Gu, Shen, & Cao, 2013).

Cost equations for desalination technologies

The total unit cost of conventional desalination *TC* is a function of capital costs *K*, non-fuel operating costs *V*, and fuel operating costs:

$$TC = \frac{K_D}{365 \cdot d_D} + \left(V_D \cdot \frac{L_D}{d_D} \right) + (E_e \cdot C_e + E_t \cdot C_f) \tag{1}$$

where K_D is the capital cost of the desalination plant (USD/m^3 per day), V_D is the non-fuel variable operation and maintenance costs (USD/m^3), L_D is the plant lifetime in years, E_e and E_t are electric and thermal energy consumption rates (kWh/m^3 and MBTU/m^3), respectively, and C_e and C_f are the costs of electricity and fuel (USD/kWh and USD/MBTU), respectively. A discount, d_D, is attributed to the desalination plant's capital and variable cost lifetime. It is calculated using an estimated amortization rate, *i*:

$$d_D = \frac{i(1+i)^L}{(1+i)^L - 1} \tag{2}$$

The unit costs of solar-powered desalination include the capital and non-energy operating costs above. Energy (fuel and electricity) costs are replaced by the per unit capital and operating costs of solar power $S_{K,V}$. The unit cost of electricity storage $B_{K,V}$ and water storage with spare desalination capacity $W_{K,V}$ are also added to the total unit costs in the appropriate scenarios. In the equations below, the capital costs, variable costs, plant lifetimes and discount rates are labelled with respect to the corresponding technology: solar (*S*), battery (*B*), or water storage (*W*):

$$S_{K,V} = \frac{1}{365 * 24} \cdot \frac{K_S}{d_S} \cdot \frac{E_e}{CF} + V_S \cdot \frac{L_S}{d_S} \cdot E_e \tag{3}$$

where K_S and V_S are the per unit capital (USD/MW) and variable cost (USD/MWh), respectively, of the solar technology, and CF is the capacity factor of the solar technology.

$$B_{K,V} = \left(\frac{K_B}{d_B} \cdot \frac{L_B}{N \cdot \varepsilon \cdot \tau} + V_B \cdot \frac{L_B}{d_B} \right) \cdot E_e \cdot (1 - CF) \quad (4)$$

where K_B and V_B are the capital (USD/MW) and variable costs (USD/MWh) of electric battery storage, τ is the discharge period in hours, ε is the discharge efficiency, and N is the number of discharge cycles.

$$W_{K,V} = \frac{K_W}{365 \cdot d_W} \cdot (1 - CF) + V_W \cdot \frac{L_W}{d_W} + \left(\frac{K_D}{365 \cdot d_D} \right) \left(\frac{1}{CF} - 1 \right) \quad (5)$$

where K_W and V_W are the capital and variable costs of water storage (USD/m^3). The last term in Equation (5) represents the capital cost associated with constructing spare desalination capacity.

Energy costs play a significant role in desalination, particularly in thermal processes, where they can constitute more than half of total costs. These costs will vary significantly between countries, depending on the endowment of energy resources and the regulations placed on prices by governments: some countries tax energy, while others sell it at prices below international market values. As stated above, in cases like Saudi Arabia, where fuel and natural gas are sold domestically at administered prices below international market values, domestic use comes at an opportunity cost. This opportunity cost will fluctuate with changes in international energy prices: when international fuel prices are high, the opportunity cost of consuming fuel domestically at administered prices is also high. The unit cost calculator takes into account the opportunity cost associated with the value of consumed energy. Specifically, the calculator investigates what opportunity cost of fuel would be necessary to balance the cost of renewable desalination options with incumbent technologies.

Finally, environmental carbon costs *(CC)* can be added to fossil-fuel-consuming technologies:

$$CC = \left(GHG_t \cdot E_t + GHG_e \cdot E_e \right) C_{GHG} \quad (6)$$

where GHG_t and GHG_e are the greenhouse gas emissions associated with thermal and electrical energy consumption (ton CO_2/MWh), respectively, and C_{GHG} is the unit cost of the emissions (USD/ton).

Although not explicitly included in this cost calculator, the land costs associated with solar power plants can be calculated separately, since they typically require a large area. In Saudi Arabia industrial land prices are typically around USD 0.02/m^2 per year, which is considered part of the fixed operating cost for solar technologies (Royal Commission for Jubail & Yanbu, 2014). Assuming that a solar power plant requires roughly 5 acres per GWH per year, land costs would contribute roughly USD 0.002/m^3, or about 0.1% of the total water cost.

Results for Saudi Arabia

The results below offer insight into the economic costs of current desalination in Saudi Arabia, as well as the prospects for adopting solar-powered technologies (see Tables 2, 3 and 4 for all figures and assumptions). These results should be interpreted as representative

Table 2. Capital and operating costs of conventional desalination technologies.

		RO on-grid	MED cogeneration	MSF cogeneration
	Capacity (m^3/day)	100,000	1,000,000	1,000,000
	Capacity factor (%)	90	90	90
L_D	Plant lifetime	25	30	30
K_D	Plant capital cost (USD/m^3 per day)	1100	1250	1500
V_D	Plant variable O&M cost (USD/m^3)	0.26	0.16	0.19
	Conventional energy			
E_e	Electrical energy (kWh/m^3)	5	2	4.25
	Thermal energy (MJ/m^3)	-	226.7	256
E_t	Thermal energy extracted for cogeneration (kWh/m^3)*	-	12	16
	Power-to-water ratio (MW/MIGD)	-	10	16

* Electrical equivalent energy divided by 50%, which represents the estimated efficiency of a stand-alone plant

Sources: Capital costs of thermal desalination plants, from UNESCO Centre for Membrane Science and Technology, University of New South Wales (2008). Capital costs of RO plants, from a project statement for the Shuaiba Expansion Project (ACWA 2014). Operating costs of all plants were extracted from Thye (2010). Thermal energy from an MED cogeneration plant based on two reported top brine temperatures in Abdel-Jawad (2001).

Table 3. Additional variables used to calculate costs of conventional and solar desalination technologies.

	Additional variables	
i	Discount rate of capital and operation costs	6%
	Administered fuel prices for Saudi Arabia	
C_t	Crude oil price for water and power utilities (2011 value)	USD 4.24/barrel
	Gas price for utilities	USD 0.75/MBTU
C_e	Industrial electricity tariff set by Saudi Electricity Company (annual average over summer and winter periods)	USD 3.84/kWh
eff	Fuel efficiency of electricity supplied by the power grid	32%

Sources: Discount rate, from Matar et al. (2014). Crude oil price and gas price for utilities, from personal communication with ECRA management (2014). Industrial electricity tariff set by Saudi Electricity Company (ECRA, 2014). Efficiency of fuel consumed for RO derived from International Energy Agency (2012).

cost estimates; actual costs will differ between projects, requiring more detailed technical analysis. As stated, the costs and operational parameters for water desalination can vary significantly, depending on plant configuration, regional labour costs, and energy. Values have been selected to represent recent desalination infrastructure projects in Saudi Arabia.

Under the current low administered energy prices in Saudi Arabia, the variable energy costs of MED represent around 23% of the operational costs. Using international fuel prices would cause energy to represent roughly 70% of operational costs. As seen in Figure 1, thermal technologies such as MED and MSF have a significant competitive advantage over RO, which has a lower dependency on primary fuel cost. When considering only administered prices (black bars), MSF cogeneration (at USD 0.85/m^3) and MED cogeneration (at USD 0.78/m^3) are about 8% and 15% cheaper than RO (at USD 0.92/m^3), respectively. The higher unit costs of the solar technologies reflect the high capital cost of investing in electricity storage or excess desalination capacity and water storage. Figure 1 shows that electricity storage is more expensive than water storage with spare desalination capacity, due to the high cost of sodium sulphur flow batteries.

Table 4. Capital and operating costs of solar desalination technologies.

		RO solar PV with electric storage	RO solar PV with water storage	RO solar CSP with water storage
	Capacity	100,000	100,000	100,000
	Capacity factor (%)	22	22	40
L_D	Plant lifetime	25	25	25
K_D	Plant capital cost (USD/m³)	1100	1100	1100
V_D	Plant variable O&M cost (USD/m³)	0.26	0.26	0.26
$S_{K,V}$	Solar technology costs			
K_S	Capital cost (USD/kW)	2500	2500	7700
V_S	Variable O&M cost (USD/kWh)	0.0034	0.0034	0.011
L_S	Technology lifetime	25	25	25
CF	Capacity factor (%)	22	22	40
$B_{K,V}$	Electricity storage costs			
K_B	Battery capital cost (USD/kW)	6100	-	-
V_B	Battery O&M cost (USD/kWh)	0.01	-	-
L_B	Lifetime	15	-	-
ε	Battery charge discharge efficiency (%)	86	-	-
τ	Discharge period (hours)	7.2	-	-
N	Discharge cycles (over lifetime)	4500	-	-
$W_{K,V}$	Water storage costs			
K_W	Capital costs (USD/m³)	-	200	200
V_W	Variable O&M costs (USD/m³)	-	0.02	0.02
L_W	Lifetime	-	30	30
CC	Carbon cost			
GHG_t	Thermal energy CO_2 emissions (g/kWh)	202		
GHG_e	Electrical energy CO_2 gas emissions (g/kWh)	630		
C_{GHG}	Cost of CO_2 gas emissions (USD/ton)	40		

Sources: Capital costs of RO plants come from a project statement for the Shuaiba Expansion Project (ACWA, 2014). Operating costs of all plants were extracted from Thye (2010). Solar PV and CSP plant costs and average yearly capacity factors from World Energy Outlook 2013 (IEA, 2014) for Middle Eastern countries. Electricity storage cost is derived for sodium sulphur (NaS) flow batteries (Akhil et al., 2013). Capital costs of water storage, from an online news article (Arab News, 2014). All costs have been adjusted to 2014 real USD.

The lower costs of thermal technologies under administered prices are unsurprising. First, the non-fuel operating expenses of thermal technologies are roughly 38% lower than membrane technologies. Second, although the energy requirements of RO are low, the primary energy must be converted to electricity, which is more capital-intensive than simply using the thermal energy from fuel. This also explains why the solar RO scenarios, which have no variable fuel component and high capital costs, are more expensive under the current fuel pricing structure.

Effectively, under these conditions, thermal desalination represents a substitution effect away from more expensive non-fuel operating costs towards relatively inexpensive fuel. Given the low administered price of fuel in Saudi Arabia, this substitution effect offers a significant cost reduction for producers.

Grey bars have been added to Figure 1 to represent the opportunity cost of consuming fuel in the three baseline scenarios: RO grid-powered, MSF cogeneration and MED cogeneration. The opportunity costs were calculated assuming a fuel price of USD 70/boe. In reality, water utilities use both natural gas and crude oil to power desalination in Saudi Arabia. Gas is more prevalent in the east, while crude oil is predominant in western provinces. When considering opportunity costs, the cost of adopting MSF cogeneration rises to USD 2.30/m³, making it

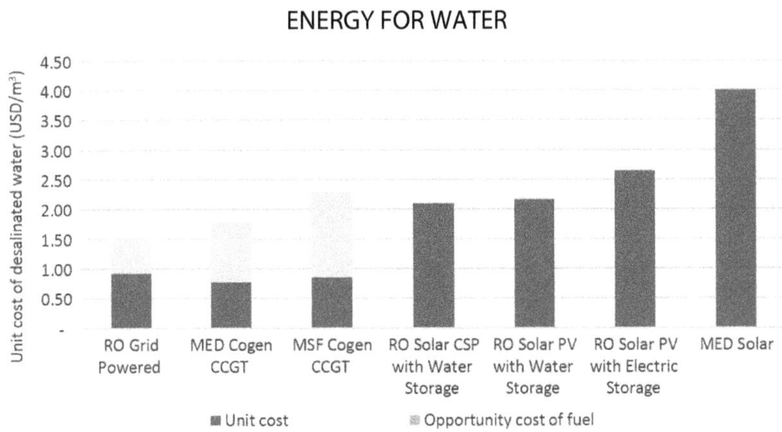

Figure 1. Estimated costs of desalination.

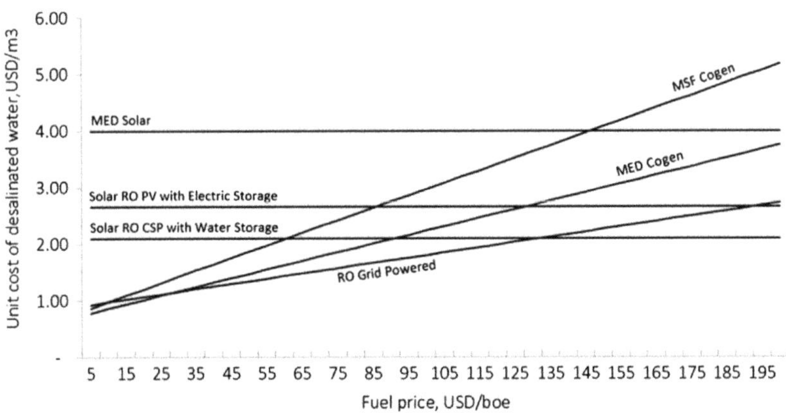

Figure 2. Relationship between energy prices and total desalination costs.

more expensive than both solar options with water storage. By contrast, MED cogeneration (USD 1.78/m^3) remains cheaper than each of the solar-powered options, but becomes more expensive than desalination using RO grid-powered (USD 1.53/m^3).

While this model offers insight into the cost of different desalination technologies in Saudi Arabia, these costs are determined using a static international fuel price of USD 70/boe. Because energy prices can change dramatically, and represent a significant portion of a desalination plant's costs, it is useful to look at how the evolution of energy prices could affect the relative costs of each scenario. Figure 2 offers this analysis, comparing the total costs of selected scenarios as the price of fuel rises.

Figure 2 highlights three points. First, as expected, the total cost of desalination powered completely by solar energy (either PV or CSP) does not vary with changes in energy prices. Second, given that thermal technologies use the most energy for desalination, the costs of MSF cogeneration and MED cogeneration are the most variable with changes in energy prices. When energy prices are below USD 30/boe, MED cogeneration offers the cheapest

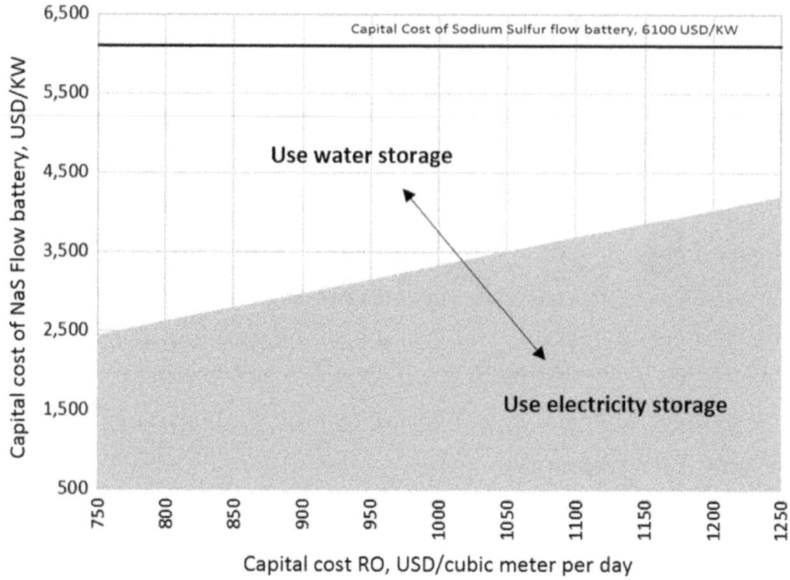

Figure 3. Breakeven cost boundary for two different solar PV desalination approaches.

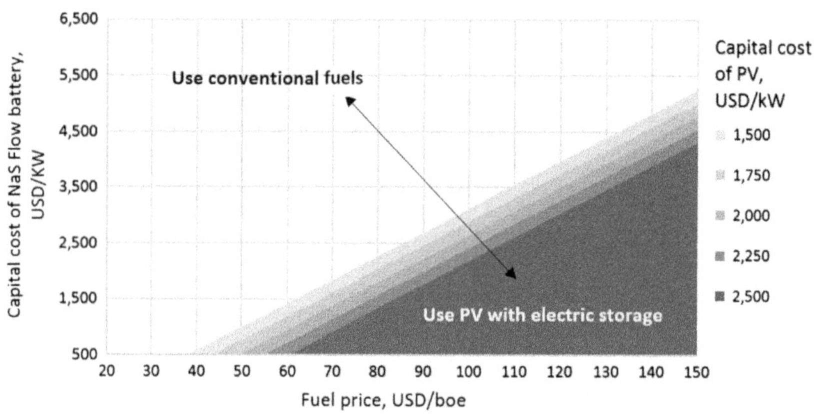

Figure 4. Breakeven cost conditions of solar PV with electricity storage versus grid-powered RO. Note: Capex = capital expenditure.

option for desalination of any technology. But MSF and MED cogeneration quickly become more expensive than solar-powered RO scenarios, exceeding the least expensive solar option once fuel prices rise above USD 60/boe and USD 90/boe, respectively.

The carbon-saving potential was also considered when assessing the breakeven price of the conventional and renewable desalination scenarios. Considering a carbon price of USD 40/ton, which is in line with the social cost of carbon estimated by the US government (Shelanski, 2013), the unit cost of RO grid-powered increases by USD 0.13/m^3, while its breakeven price with the cheapest solar option, CSP with water storage, is reduced to

USD 118/boe. Therefore, without extremely high fuel prices or a very high price on carbon, environmental costs do not sufficiently improve the competitiveness of solar desalination.

Additional breakeven analyses were performed to determine how technological innovation in energy storage, spare desalination capacity and solar power could improve the competitiveness of solar desalination compared to conventional RO. Specifically, the unit cost calculator identifies what opportunity cost of fuel would be necessary to balance the cost of renewable desalination options with incumbent technologies. This breakeven cost depends on the capital and operating costs consistent with each scenario. The article uses a simple linear cost function $f_i(X)$, see Equation(6), defined for each scenario i, depending on a vector of input cost variables ($X = x_1, x_2, \ldots$), such as the opportunity cost of fuel and technology costs, with coefficient vector A_i, and constant cost coefficient C_i:

$$f_i(X) = A_i \cdot X + C_i \tag{6}$$

Taking RO grid-powered as a baseline scenario b, $f_b(x_1, x_2)$, and equating it with a renewable scenario r, such as RO solar PV with water storage, $f_r(x_1, x_2)$, identifies the set of input variables $\{x_1^*, x_2^*\}$ where the costs of the two scenarios break even. For example, the cost boundary intersection between these two scenarios can be determined for a given breakeven desalination capital cost, K_D^*, as a function of the breakeven fuel cost C_f^* (Equation (7)). The term added at the end, f_b, accounts for the cost of fuel consumed for desalination in excess of the actual cost paid by the power utility, C_f. The coefficient *eff* represents the fuel consumption efficiency of the electric grid.

$$\begin{gathered} f_r(K_D) = \frac{K_D}{365 \cdot d} \cdot \frac{1}{CF} + V_D \cdot \frac{L_D}{d_D} \\ f_b(K_D, C_e) = \frac{K_D}{365 \cdot d} + V_D \cdot \frac{L_D}{d_D} + E_e \cdot C_e + \frac{E_e}{eff} \cdot (C_f^* - C_f) \\ f_r(K_D) - f_b(K_D, C_e) = 0 \rightarrow \\ K_D^*(C_f^*) = \left(\frac{E_e}{365 \cdot d} \cdot \frac{CF}{1-CF}\right) \cdot \left(C_e + \frac{1}{eff}(C_f^* - C_f)\right) - \left(W_k \cdot CF \cdot \frac{d}{d_w}\right) \end{gathered} \tag{7}$$

First, consider the trade-off between desalination powered by solar PV using either electric storage or excess desalination capacity with water storage. The breakeven conditions between these two scenarios, including the levellized cost of electricity storage (or capital cost of electric batteries) and cost of spare desalination capacity, are considered in Figure 3. Given present technologies, electricity storage using sodium sulphur flow batteries is simply too expensive compared to the cost of spare desalination capacity (USD 900–1200/m^3 per day). Under the assumptions of this article, the levellized cost of the bulk sodium sulphur batteries, accounting for a 6% discount rate, is roughly USD 0.34/kWh. The levellized cost of electric storage would have to drop well below USD 0.25/kWh before this technology could be competitive with water storage (given current capital costs).

The conditions necessary to reach the breakeven cost between RO solar with electric storage and conventional RO are illustrated in Figure 4. The variables considered are capital cost of battery storage (vertical axis), fuel cost (horizontal axis), and capital cost of solar power (shaded regions). The shaded regions identify where the unit cost of water is lower for the solar option. The figure shows that for current capital costs of sodium sulphur flow batteries (USD 6100/KW), and solar PV costs in the range of USD 2000/KW, fuel costs would have to exceed USD 150/boe for solar to break even. However, that breakeven oil price drops into the USD 70/boe range if the capital cost of battery storage falls to one-quarter (USD 1500/KW).

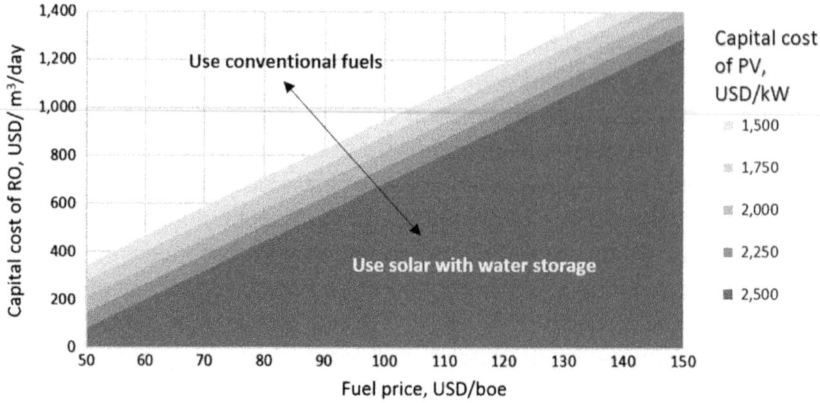

Figure 5. Breakeven cost conditions of solar PV with water storage versus grid-powered RO.

Last, the conditions in which Scenario 3 (RO solar PV with water storage) could compete with conventional RO are considered. The breakeven cost boundary between RO solar with water storage and conventional RO are illustrated in Figure 5. It should be noted that water storage costs represent less than 3% of total costs, and therefore do not play a major role in the relative economics. Assuming that innovation reduced the capital cost of PV and RO to USD 2000/MW and USD 800/m^3 per day respectively, solar desalination with water storage could become cost-competitive with energy costs of roughly USD 100/boe. This breakeven analysis was made assuming a 5 kWh/m^3 electricity consumption rate for RO. For more energy-intensive membrane treatment, which can be the case for the high-salinity Red Sea and Arabian Gulf waters, the breakeven will be reduced, thus further increasing the competitiveness of solar desalination.

The results are dependent on the discount rate assigned to these projects. A high discount rate disproportionally increases the solar desalination costs, as they are highly capital-intensive compared to conventional systems. Conversely, a low discount rate would improve the cost-competitiveness of solar technologies.

Conclusions and policy implications

As the population and economy of Saudi Arabia grow, so will demand for desalinated water. This article has offered a framework for estimating the costs of different solar desalination options compared to incumbent technologies. Here, *solar desalination* is defined as a desalination plant that obtains solar energy from a closed system. This was done to isolate the investigation to issues relevant for desalination applications, rather than studying the efficacy of replacing conventional energy sources with renewables in an integrated electricity grid.

The results demonstrate that Saudi Arabia's current strategy of using thermal desalination technologies makes economic sense only with the current administered prices of fuels. Raising fuel prices to market levels would incentivize a shift to more energy-efficient RO, reducing the total primary energy consumed for desalination. In fact, a fuel price above USD 35/boe provides the necessary condition to switch from energy-intensive thermal cogeneration to more energy-efficient RO.

The prospects for adopting solar-powered energy for desalination given the costs of current technologies, however, are less clear. As stated, integration of state-of-the-art solar power and membrane desalination technology is currently being pursued in Saudi Arabia at the Al-Khafji solar desalination plant. However, under the technical and cost assumptions presented in this article, such systems are unlikely to be cost-competitive with conventional desalination.

Given current technologies, adopting the cheapest solar desalination option, RO solar CSP with water storage, would increase the total costs of water to USD 2.10/m^3. This is the equivalent of desalinating water using RO grid-powered when fuel prices are in the range of USD 132/boe. Should a transition to solar power be adopted, the increased costs would probably be borne by producers, given that consumers in Saudi Arabia are sensitive to utility price increases. Increasing costs to producers, however, would place further strain on the finances of water utilities. Under the current cost structure, total annual revenues do not exceed 2.5% of annual expenditure (Abderrahman, 2007), with most revenues coming from the industrial sector, as residents pay less than USD 0.05/m^3 (Zetland, 2014). For this reason, any shift to more expensive solar desalination technologies is likely to be politically sensitive.

While solar desalination options are presently uncompetitive, innovation is likely to bring down costs. Presently, the cost of bulk flow batteries is very high, so without a major technological breakthrough it is unlikely that electricity storage will be competitive compared to alternative technologies. However, some advances have been made in this area. For example, compressed-air energy storage, in which air is compressed and stored in an underground cavern and released to power a turbine, provides a low-cost energy storage option. Studies suggest that compressed-air energy storage can achieve levellized electricity storage costs in the range of USD 0.12 kWh. At this cost, RO solar PV with electric storage could break even with conventional RO at a fuel cost of roughly USD 80/boe. But this technology is geographically constrained to areas with access to both well-characterized subsurface aquifers and saline water supplies, so it is not presently considered a viable option on a large scale.

A more promising development is the cost of spare desalination capacity, which has fallen significantly in recent years. Should this trend continue, solar-powered desalination with excess capacity and water storage could become competitive with incumbent technologies.

Although stand-alone solar thermal desalination is currently uneconomic, at USD 4/m^3, other low-cost methods of harvesting heat could be considered, such as complementing the waste heat from a standard cogeneration power plant with solar thermal energy. The low-cost solar thermal collector could be used to reduce the heat extracted for the desalination process, increasing the overall efficiency of the power plant.

This article has limited its analysis to solar-powered technologies. There are, however, other unconventional renewable options for desalination. One example is the use of wind energy, which can complement solar technologies by supplying additional power after sunset. Another option is using low-quality geothermal energy for low-temperature desalination of brackish water. Employing the methodology of this article, further study should be done in these areas to assess the cost-competitiveness of these technologies (both in isolation and in conjunction) as energy sources for desalination.

Disclosure statement

No potential conflict of interest was reported by the authors.

References

Abdel-Jawad, M. (2001). *Energy sources for coupling with desalination plants in the GCC countries.* ESCWA Consultancy Report.

Abderrahman, W. A. (2007). Water management in ArRiyadh. *International Journal of Water Resources Development, 22*, 277–289. doi:10.1080/07900620600654785

Abderrahman, W. A. (2010). Energy and water in Arid developing countries: Saudi Arabia, a case-study. *International Journal of Water Resources Development, 17*, 247–255. doi:10.1080/07900620120031306

Abengoa. (2015). Abengoa and AWT to develop the world's first solar-powered desalination plant, in Saudi Arabia. Retrieved from Abengoa website: http://www.abengoa.com/web/en/noticias_y_publicaciones/noticias/historico/2015/01_enero/abg_20150121.html

ACWA Power. (2014). *Project statement for the Shuaiba expansion project.* Retrieved from ACWA Power website: http://www.acwapower.com/project/4/shuaibah-expansion-project-company.html

Akhil, A. A., Huff, G., Currier, A. B., Kaun, B. C., Rastler, D. M., Chen, S. B., Cotter, A. L., Bradshaw, D. T., & Gauntlett, W. D. (2013). *DOE/EPRI 2013 Electricity Storage Handbook in Collaboration with NRECA.* (Report No. SAND2013-5131). Albuquerque, NM: Sandia National Laboratories.

Al-Ibrahim, A. M. (2013). Seawater desalination: The strategic choice for Saudi Arabia. *Desalination and Water Treatment, 51*, 1–4. doi:10.1080/19443994.2012.704744

Aquastat. (2014). *Saudi Arabia fact sheet, 2008.* Retrieved from the Food and Agriculture Organization of the United Nations: http://www.fao.org/nr/water/aquastat/countries_regions/sau/index.stm

Arab News. (2014, February 24). Strategic water storage facility to cost SR1bn. Retrieved from http://www.arabnews.com/news/530411?quicktabs_stat2=1

Bernat, X., Gibert, O., Guiu, R., Tobella, J., & Campos, C. (2010). The economics of desalination for various uses. In Martínez-Cortina, L., Garrido A., and López-Gunn, E. (Eds.), *Re-thinking water and food security* (pp. 329-346) CRC Press. doi: 10.1201/b10541-25

Daya, A., & El Baltaji, D. (2012). Saudi Arabia may become oil importer by 2030, citigroup says. *Bloomberg.* Retrieved from http://www.bloomberg.com/news/2012-09-04/saudi-arabia-may-become-oil-importer-by-2030-citigroup-says-1-.html

Desal Database. (2014). Published by Global Water Intelligence. Retrieved from http://desaldata.com/

ECRA. (2014). *Activities and achievements of the authority in 2013.* Retrieved from ECRA website: http://www.ecra.gov.sa/documents/Annual%20Reports/Annualreport%20for%20Web2013-english.pdf

Elhadj, E. (2004). *Camels don't fly, deserts don't bloom: An assessment of Saudi Arabia's experiment in desert agriculture.* Occasional Paper No 48, Water Issues Study Group, School of Oriental and African Studies (SOAS)/King's College London.

Gately, D., Al-Yousef, N., & Al-Sheikh, H. M. H. (2012). The rapid growth of domestic oil consumption in Saudi Arabia and the opportunity cost of oil exports foregone. *Energy Policy*, Vol. 47, August, 57-68. doi: 10.1016/j.enpol.2012.04.011

Giansiracusa, A. (2010). "Coping with scarcity: Saudi Arabia and water", *Stimson Center Spotlight.* Retrieved from http://www.stimson.org/spotlight/coping-with-scarcity-saudi-arabia-and-water/

Gude, V. G., Nirmalakhandan, N., Deng, S., & Maganti, A. (2012). Low temperature desalination using solar collectors augmented by thermal energy storage. *Applied Energy, 91*, 466–474. doi:10.1016/j.apenergy.2011.10.018

International Energy Agency. (2012). Saudi Arabia: Balances for 2012. Retrieved June 15, 2015, from http://www.iea.org/statistics/statisticssearch/report/?year=2012&country=SAUDIARABI&product=Balances

International Energy Agency. (2014). World energy outlook 2013 power generation in the new policies and 450 scenarios report. Retrieved June 2, 2015, from http://www.worldenergyoutlook.org/weomodel/investmentcosts/

Kalogirou, S. A. (2005). Seawater desalination using renewable energy sources. *Progress in Energy and Combustion Science, 31*, 242–281. doi:10.1016/j.pecs.2005.03.001

Liu, X., Chen, W., Gu, M., Shen, S., & Cao, G. (2013). Thermal and economic analyses of solar desalination system with evacuated tube collector. *Solar Energy, 93*, 144–150. doi:10.1016/j.solener.2013.03.009

Matar, W., Murphy, F., Pierru, A., & Rioux, B. (2014). Lowering Saudi Arabia's Fuel consumption and energy system costs without increasing end consumer prices. *Energy Economics* 49, 558–569. doi:10.1016/j.eneco.2015.03.019 .

Ministry of Water & Electricity. (2013). *Annual report 2012*. Retrieved from Ministry of Water & Electricity website: http://www.mowe.gov.sa/MOWE2012AnualReport/electricity2012.html#/69/

Ouda, K. M. O. (2013). Water demand versus supply in Saudi Arabia: Current and future challenges. *International Journal of Water Resources Development, 30*, 335–344. doi:10.1080/07900627.2013.837363

Qiblawey, H.M., & Banat, F. (2008). Solar thermal desalination technologies. *Desalination, 220*, 633–644.

Royal Commission for Jubail and Yanbu. (2014). Why Saudi Arabia, economy and resources. Retrieved from: http://www.rcjy.gov.sa/en-US/AboutUs/WhySaudiArabia/Pages/default.aspx

Saif, O. (2012). *The future outlook of desalination in the Gulf: challenges & opportunities faced by Qatar & the UAE*. Environment and Health Research Paper: United Nations University for Water.

Shelanski, H. (2013). Refining estimates of the social cost of carbon. Retrieved at: https://www.whitehouse.gov/blog/2013/11/01/refining-estimates-social-cost-carbon

Thye, J. F. (2010). *Desalination: Can it be greenhouse gas free and cost competitive?* (Unpublished master's thesis). New Haven, CT: Yale School of Forestry and Environmental Studies.

UNESCO Centre for Membrane Science and Technology, University of New South Wales. (2008, October). *Emerging trends in desalination: A review*. (Waterlines Report Series No 9). Canberra: National Water Commission.

World Bank. (2012). *Renewable energy desalination. An emerging solution to close the water gap in the middle East and North Africa, MENA development report*. Washington, D. C.: The World Bank.

World Bank. (2014a). Indicators - population, total. Retrieved May 7, 2015, from http://data.worldbank.org/indicator/SP.POP.TOTL/countries/SA?display=default

World Bank. (2014b). Indicators - GDP per capita (current US$). Retrieved May 7, 2015, from http://data.worldbank.org/indicator/NY.GDP.PCAP.CD/countries/SA?display=graph

Zetland, D. (2014). Rational waste: The political-economy of desalination. *Leiden University Discussion Paper*.

Responding to water challenges in Greece through desalination: energy considerations

A. Kartalidis[a], E. Tzen[b], E. Kampragkou[a] and D. Assimacopoulos[a]

[a]School of Chemical Engineering, Technical University of Athens, Athens, Greece; [b]Independent Engineer, Athens, Greece

ABSTRACT
Desalination technology, and reverse osmosis in particular, is used by several island authorities in Greece to address water scarcity. However, this is a highly energy-intensive technique, requiring the consumption of significant quantities of fossil fuels. The case of Syros Island is presented, to demonstrate the strong water–energy link in the operation of desalination plants. The article also discusses the use of renewable energy sources as a means for reducing the energy intensity of desalination.

Introduction

Economic development and social well-being depend on two vital resources, energy and water, the links between which are both complex and significant: energy production requires large quantities of water, and supplying water requires significant amounts of energy. The analysis of the "water–energy nexus" demands an integrated view of the production and consumption chains of each resource (Marsh, 2008, presents a depiction of the water–energy nexus complexity), focusing on the core operational components, i.e. infrastructure and technologies (Scott et al., 2011).

Particularly for the water supply chain, there is a pressing need to map energy demand along the stages, as the energy intensity of each stage depends on the input water quantity and quality. Overall, the stages of the water supply chain tend to become more energy-demanding over time (Schumacher, 2007). Energy intensity can increase with growth in water demand, extraction of water from deeper bores, longer interbasin water transfers, stricter water quality standards, and greater restrictions on wastewater discharge. Despite all this, there are options for improving energy efficiency. For example, as highlighted by the European Innovation Partnership for Water (EIPW, 2015), there is significant potential for increasing energy use efficiency in water supply systems by introducing low-energy technologies in water and wastewater processes and increasing the use or efficiency of renewable resources.

The energy intensity of a water supply chain can vary depending on the sources of water (type, seasonal availability, quality) and end-user characteristics (infrastructure, consumption patterns). The anticipated impacts of climate change on water resources (IPCC, 2014), the increased competition among users, and the "loop effect" of the energy–water interlinks are expected to lead to a 35% increase in energy consumption by 2035. This could in turn lead to an 85% increase in water consumption and thus energy requirements for water treatment and distribution (International Energy Agency, 2012; Rodriguez, Delgado, DeLaquil, & Sohns, 2013). Given these interlinkages, it is important to analyze the energy–water relationship, particularly in water-scarce regions.

In water-scarce regions, three main supply management approaches are followed: (1) transferring water from areas of relative abundance to areas of relative scarcity; (2) intensive use of available resources; and (3) use of alternative water resources (e.g. desalination, wastewater reuse). All three result in an increase in the energy intensity of the water system, leading to high cost of water, either for consumers or governments. Thus, any improvements in the efficiency of water and energy use may lead to benefits for the environment as well as reductions in the cost of water provision paid by consumers or governments.

This article discusses the energy-for-water links in Greece, focusing on the most water-scarce regions of the country, the arid and semi-arid islands of the Aegean Sea. On these islands, desalination currently offers the most effective solution for meeting public water demand, though it is one of the most energy-intensive methods (together with transport and conveyance). Syros Island, located at the centre of the Cyclades island complex, is used as a case study to demonstrate the links between water production and energy consumption, and the need to increase energy use efficiency in the isolated water systems of the Greek Islands.

The water system of the Greek islands: energy considerations

The problem of scarcity

Greece is characterized by the longest coastline in Europe, at roughly 14,000 km. The country has about 2,500 islands (see Figure 1), with a total area of 21,580 km^2. The population of the islands is estimated at 1,633,433 (EL.STAT., 2012).

The climate on the Greek islands is Mediterranean (wet winters and dry summers), with low annual precipitation (less than 400 mm, especially for the Aegean Islands; Hellenic National Meteorological Service, 2015). The low precipitation, in addition to the geological formations, makes the storage of surface or groundwater of sufficient quantity and quality impossible on some islands. Moreover, the seasonal summer peak in water demand (due to irrigation and tourism), the high water losses due to leakages in old distribution networks, the absence of integrated water management systems and the degradation of aquifers due to over-pumping all intensify water-scarcity problems and tend to create local water deficiencies (Table 1).

Water deficiency problems are addressed in different ways according to the special characteristics of each island. The current practices for domestic water supply, according to Nokas (2002), are: (1) desalination of sea or brackish water using reverse osmosis (RO) and/or thermal technologies; (2) water transfer from the mainland or nearby islands on tanker ships; (3) large reservoirs, dams and tank-lakes that store surface and rain water; and (4) boreholes.

Figure 1. Map of Greek islands and island complexes.

Table 1. Island complexes in Greece (EL.STAT., 2012).

Complex	Number of islands	Population	Water balance (Special Secretariat for Water [SSW], 2013–14)
Dodecanese	24	191,084	Negative
Ionian	17	211,954	Positive
Cyclades	28	118,000	Negative
North Aegean	14	202,360	Negative
Sporades	7	16,792	Negative
Saronic	7	63,467	Negative

Domestic use of surface water is limited to the largest islands; the majority of the Aegean Islands use groundwater from boreholes and/or desalinated water from brackish or seawater desalination plants. In many cases, water is transferred by tankers from the mainland or from nearby islands to assist water supply. Despite being costly and unsustainable, this practice remains popular due to the unwillingness of some public officials to adopt alternative solutions such as desalination.

Desalination: an energy-intensive water source

In the 1960s, when it was understood that water transport was not cost-effective, desalination was identified as a potential tool for addressing water scarcity on the Greek islands. The first attempts at desalination involved the use of solar still collectors filled with seawater.

Heat provided by the sun evaporates the water, separating it from salt and other impurities. The vapour is condensed on an inclined glass surface at the top of the collector, and the distilled water is collected in the edge. From 1964 to 1973 five solar still systems of different sizes were installed on the islands of Symi, Kastelorizo, Kimolos, Aegina and Patmos. Their individual water production was approximately 3 L per day per m^2 of collector area. The projects failed for various reasons, including unskilled personnel, operation malfunctions due to bad design (at the early installations), the lack of maintenance programmes (Deligiannis & Belesiotis, 1995), and conflicts of interest with water tanker owners (success in the operation of desalination plants would result in significant reduction in the quantities of water they transferred).

The second attempt at using desalination was in Syros in 1969–70 with the construction of a thermal multi-stage flash unit of 1200 m^3 daily capacity. The desalination unit operated with fuel oil and was in operation until 1984, by which time several technical problems had emerged, mostly due to scaling in the heat exchangers. The technical problems, in combination with the high cost of spare parts and the cost of oil, led to the transition towards RO (Vakondios, 2015).

In 1977–78, an electrodialysis reversal plant with daily capacity of 15,000 m^3 was installed on the island of Corfu for the treatment of low-salinity brackish water (up to 2000 ppm). Its operation was halted due to functional problems after a few years (Dafnis, 2005). Its specific energy consumption was 1.7 kWh/m^3 (Arnold, 1979).

The first RO units for public water supply were installed on the islands of Ithaca and Mykonos in 1981–82. The actual specific energy consumption of the Ithaca RO plant was 15 kWh/m^3, and the installed power of the unit was 510 kW. Since these first installations, RO has been the preferred technology for water desalination; it has been used for more than 30 years for public water supply, and privately by hotels and luxury houses. The reasons for the adoption of RO are: (1) modular operation to meet seasonal variations in demand; (2) relatively low water needs; (3) compact design; (4) fast installation (2–3 months); (5) low energy consumption relative to other desalination methods; and (6) easy operation.

Presently, the total installed desalination capacity (for public use in the islands) is around 52,000 m^3/day, distributed among 30 islands (Figure 2), with 9000 m^3/day brackish water feed and 43,000 m^3/day seawater feed. The majority of the desalination units are installed on islands with autonomous electrical energy production grids (22 of the 30 islands), and the rest are on islands that are interconnected with the mainland's electrical grid through underwater cables. The rate of desalination development was quite low until the early 2000s, but rose rapidly prior to the Athens Olympic Games in 2004 (Figure 3). Around 10 islands are expected to install desalination plants, with a total capacity of 11,000 m^3, in the coming years.

The energy consumption of RO units is high and takes place almost exclusively in three pump stages (pumping water from the sea to the desalination plant; pretreatment and desalination; and product water forwarding to the distribution tank). Over the years, great advances have been made to reduce energy consumption through the use of more efficient energy-recovery devices. Today, all desalination units (even small units producing around 100 m^3/day) use efficient and reliable energy-recovery devices to significantly reduce energy use. The specific energy consumption for selected desalination units installed on the islands is presented in Figure 4 from data acquired from manufacturers and local water companies (Culligan Hellas SA, personal communication on technical characteristics of existing

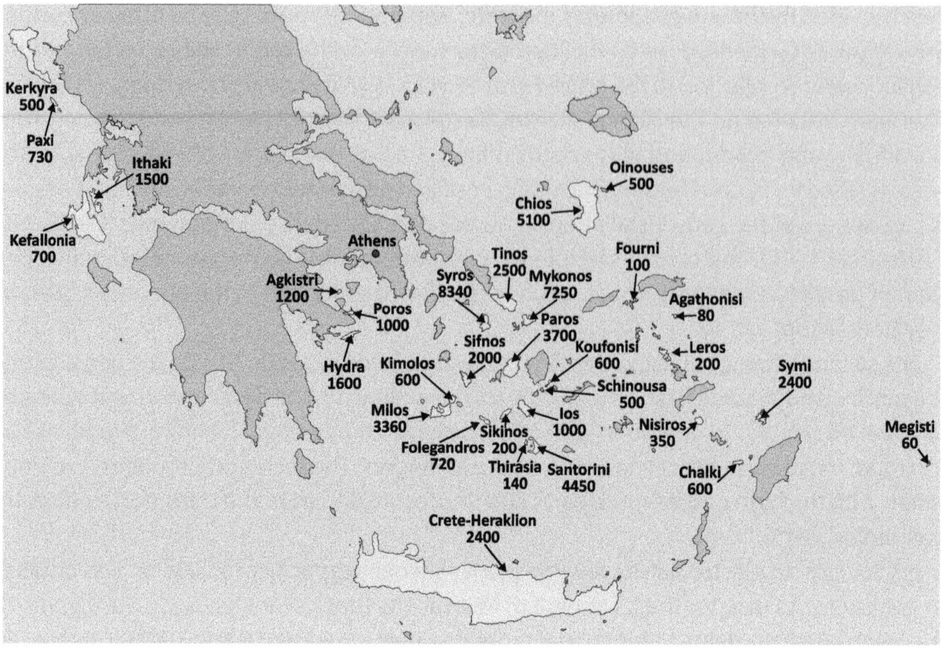

Figure 2. Desalination units in the Greek islands (Cuenca, 2012; Dagkalidis, 2009; Dalmira, 2013; imerisia.gr., 2009; Kampourakis, 2013; Karagiannis & Soldatos, 2007; sychem.gr., 2011; TEMAK SA, personal communication, 2014; Vakondios, 2015; Ventouris, 2013).

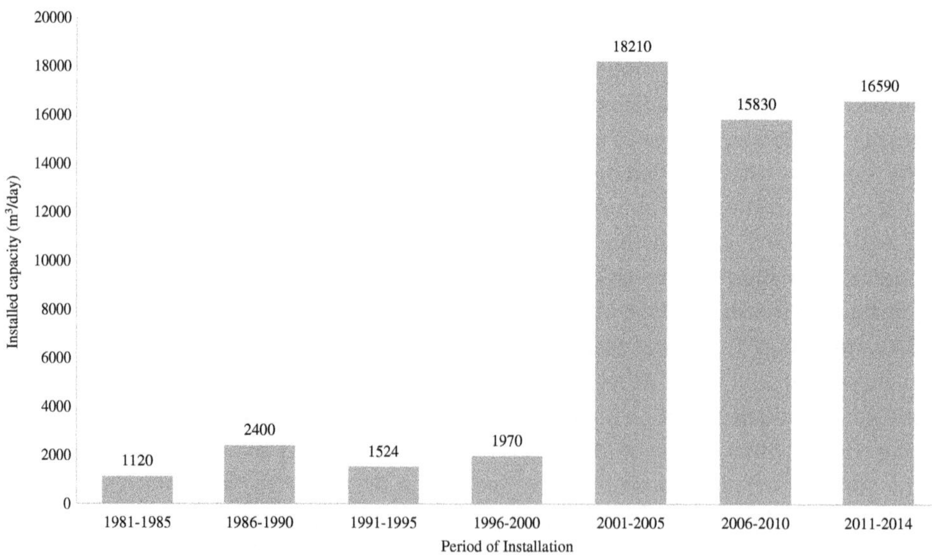

Figure 3. Desalination facilities in Greece, 1981–2014 (Cuenca, 2012; Dagkalidis, 2009; Dalmira, 2013; imerisia.gr., 2009; Kampourakis, 2013; Karagiannis & Soldatos, 2007; sychem.gr., 2011; TEMAK SA, personal communication, 2014; Vakondios, 2015; Ventouris, 2013).

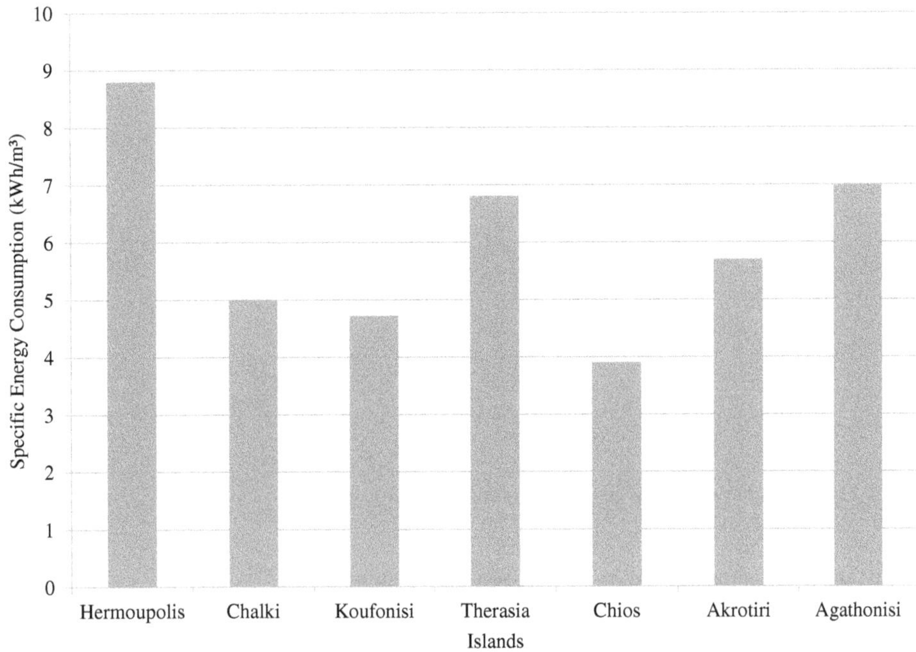

Figure 4. Specific energy consumption of selected desalination units in Greece (TEMAK SA, personal communication, 2014; Syros WSSC, personal communication, 2014).

desalination units, 2014; TEMAK SA, personal communication, 2014). Both booster and high-pressure pumps are included in the calculations.

In Chios, Koufonisi and Chalki, the specific energy consumption is estimated at less than 5 kWh/m^3. In Thirasia, Agathonisi and Akrotiri, this value is higher due to the small size of these plants (installed capacity less than 250 m^3/day), making auxiliary energy consumption relatively more important. In Hermoupolis, the capital of Syros Island, the actual recorded specific energy consumption is 9 kWh/m^3. This high value is attributed to the use of older-technology energy-recovery equipment, installed before 2002, when pressure-exchanger technology was expensive and technologically immature.

The case of Syros Island: comparing the operation and energy costs of desalination

The urban water supply in Syros consists mainly of desalinated water. The island has an autonomous grid, and the electrical energy supply relies on diesel fuel. Currently, there are 13 RO units distributed in five areas (Figure 5). Twelve of these use the mature and reliable, but also less efficient, Pelton wheel turbines as energy-recovery devices, and one uses the more advanced pressure exchanger. The total capacity of all the plants is 8340 m^3/day, with individual capacities ranging from 250 to 2000 m^3/day. According to the Hellenic Electricity Distribution Network Operator (HEDNO, 2012) and the Syros Water Company (Syros Water Supply and Sewerage Company [Syros WSSC], 2014), more than 11% of the total electricity produced in the island is used for water production, and the desalination power demand when plants are running at full capacity is 5.2% of the conventional installed power.

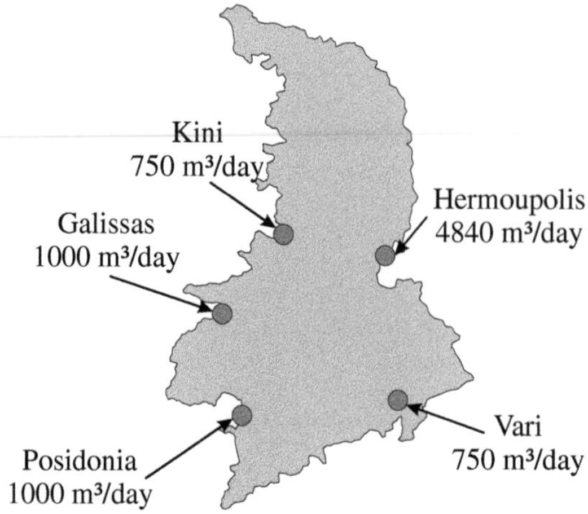

Figure 5. Location and capacity of desalination plants in Syros Island.

Table 2. Data used for the calculation of the assessment indicators.

	Hermoupolis DEYA Syrou	Athens EYDAP	Thessaloniki EYATH
Population served	13,400	4,500,000	1,000,000
Water production (hm^3)	1.05	385.5	89.7
Water billed (hm^3)	0.89	337	66.3
Annual per capita consumption (m^3/year)	66.6	74.9	66.3
Energy consumption (GWh)	10	251	119
Reference year or period	2010–13	2009	2008
Energy cost (€/kWh)	0.086	0.089	No data(assumed same as Athens)
Data sources	Syros WSSC (2014)	EYDAP, Athens Water Supply & Sewerage Company (2011); Georgalas (2010)	EYATH, Thessaloniki Water Supply & Sewerage Company (2010); Georgalas (2010)

The 13 desalination units have a total installed power of 2.08 MW (only booster and high-pressure pumps). The cost of energy is the main operating cost for the Syros Water Supply Company, which estimates the total operational cost at €1.2–1.6/m^3. The participation of energy in the cost of water is €0.7/m^3 (about 45%), with an average energy cost of €0.086/kWh (Vakondios, 2015). Energy production is subsidized in order to offer consumers an electricity rate similar to that of the mainland. Thus, the real cost of energy is higher than what is billed, and the participation of energy in the cost of water can be as high as €1.7/m^3, which is roughly 240% of the billed energy for water production.

The energy-for-water connection is described using three indicators (energy consumption per volume of water sold, energy cost per volume of water sold, and energy consumption per capita), including both the water supply and wastewater treatment processes in the analysis. The present analysis concerns only the desalination plant of Hermoupolis, for which the required data are available. For the purpose of this comparison, the same results are given for the two major water supply systems in Greece (Table 2): the Athens water supply system, which relies exclusively on fresh surface water; and the Thessaloniki water supply system, which is based on a mix of fresh groundwater and fresh surface water.

Assessment indicators

Figure 6 presents the energy consumption per volume of water sold in the three water systems (water supply and treatment). Athens has a specific energy consumption of 0.75 kWh/m^3, Thessaloniki has twice that, and Hermoupolis' consumption is 15 times that of Athens, at 11.2 kWh/m^3. The cost of energy follows a similar pattern, as shown in Figure 7. Similar ratios also apply for the case of energy demand for water supply per capita. The value in Syros is 14 times that in Athens, and more than twice that in Thessaloniki (Figure 8). The assessment indicators showcase how the source of water affects the demand for energy, with desalination plants having the highest demand.

Monthly variation of specific energy consumption

Water demand on the Island of Syros is higher during the summer period, mainly as a result of tourist influx. For example, water production in August is about 85% higher than in February

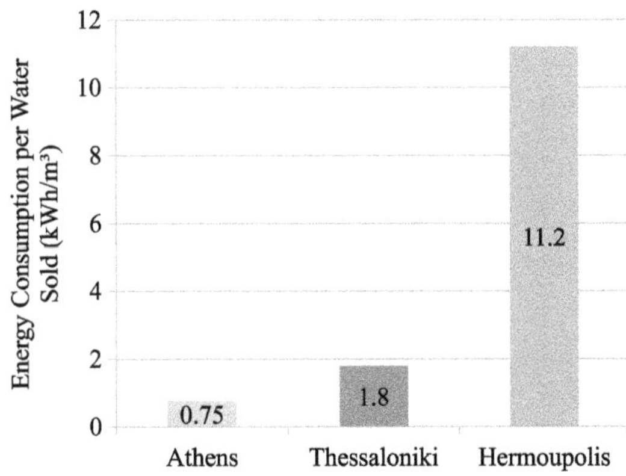

Figure 6. Energy consumption per volume of water sold in the water supply systems of Hermoupolis, Athens and Thessaloniki (Syros WSSC, 2014; EYDAP, Athens Water Supply & Sewerage Company, 2011; EYATH, Thessaloniki Water Supply & Sewerage Company, 2010; Georgalas, 2010).

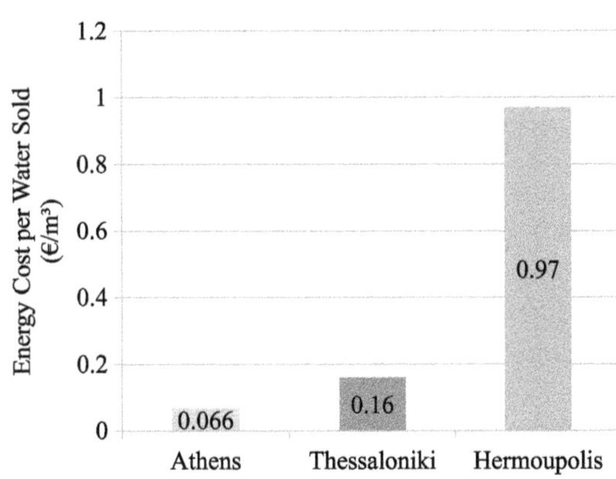

Figure 7. Energy cost per volume of water sold in the water supply systems of Hermoupolis, Athens and Thessaloniki (Syros WSSC, 2014; EYDAP, Athens Water Supply & Sewerage Company, 2011; EYATH, Thessaloniki Water Supply & Sewerage Company, 2010; Georgalas, 2010).

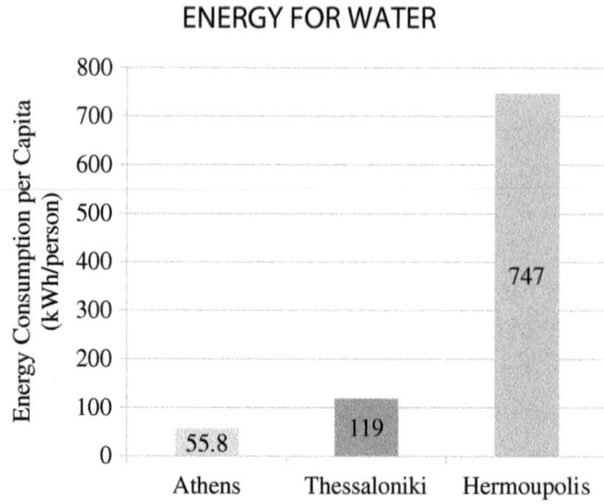

Figure 8. Energy consumption per capita in the water supply systems of Hermoupolis, Athens and Thessaloniki (Syros WSSC, 2014; EYDAP, Athens Water Supply & Sewerage Company, 2011; EYATH, Thessaloniki Water Supply & Sewerage Company, 2010; Georgalas, 2010).

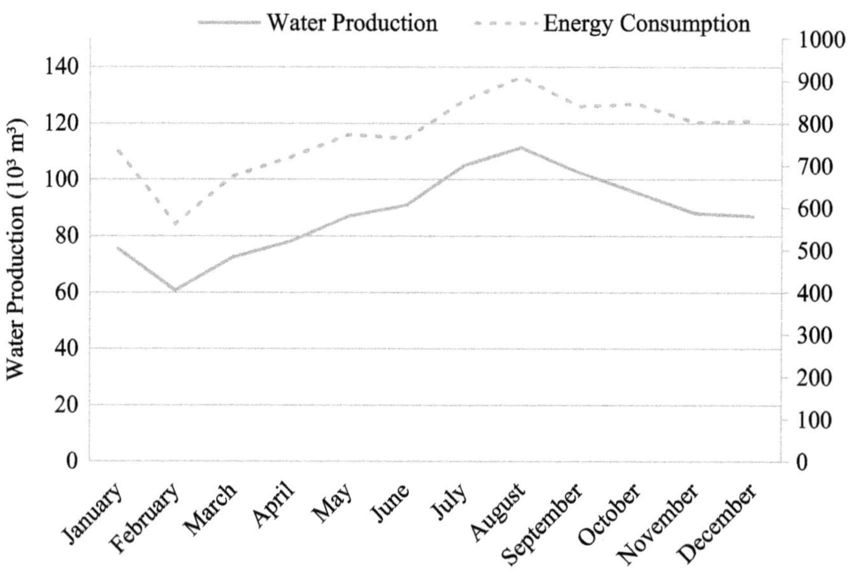

Figure 9. Monthly water production and energy consumption for the Hermoupolis desalination plant (average values for 2010–2013) (Syros WSSC, 2014).

(Figure 9). Similar variability applies to the energy consumed by the desalination plant. In February, the energy required is only 560 MWh, but in August it increases by 62.5%, to 910 MWh (Figure 9). This difference shows that the desalination process is more efficient in summer than in winter (Figure 10). Desalination specific energy for the Hermoupolis plant is about 8.1 kWh/m^3 in August and 9.7 kWh/m^3 in February, a reduction of about 16.5%. This reduction is attributed to three main factors: (1) the desalination units are working continuously during the summer period, minimizing start-and-stop cycles and consequently the energy-consuming cleaning process; (2) the higher temperature of seawater during summer changes water properties such as its viscosity, lowering the specific energy demand of the

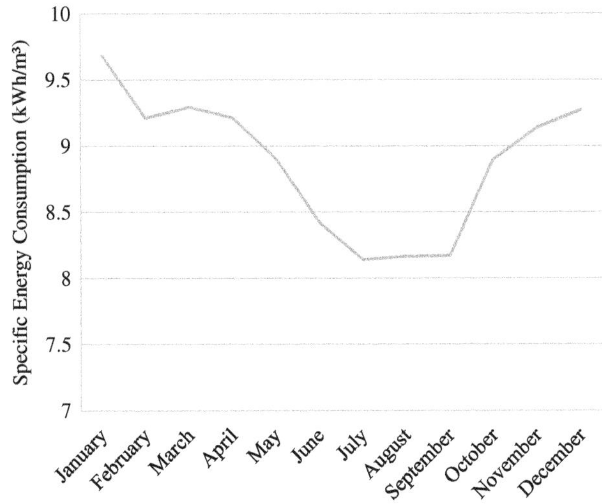

Figure 10. Specific energy consumption of the desalination units in the Hermoupolis desalination plant (average values for 2010–2013) (Syros WSSC, 2014).

high-pressure pump for the desalination process (Water Reuse Association, 2011); and (3) as the Hermoupolis desalination plant consists of several desalination units, each with different capacities and specific energies, making extensive use of the unit with the lower specific energy also lowers the specific energy of the desalination plant as a whole.

Desalination as a supply option

The per capita annual water consumption for domestic use in Hermoupolis is estimated at 66.6 m^3 (Table 2). This value is almost equal to the water consumption per capita in the city of Thessaloniki, and 12% less than in Athens. It is important, however, to note that there is a significant difference between water use rates by local inhabitants and tourists (about 120 L/day and 200 L/day, respectively), giving the 'misleading message' that even though Syros is lacking in water resources, the consumption is as high as in the urban centres of Thessaloniki and Athens. The summer peak demand increases the per capita consumption, and if the current tourism development pattern continues, new water supply resources (i.e. new desalination plants) may be required. This claim is supported by the fact that water consumption in Hermoupolis increased by 5.6% from 2009 to 2013 (Syros WSSC, 2014).

New desalination plants to meet an increasing water demand will, thus, increase energy requirements of the isolated water systems, if energy efficiency measures are not considered.

Desalination with renewable energy sources

The use of renewable energy for desalination offers significant benefits in comparison to conventional energy systems. Although renewable energy desalination systems have higher installation costs, their operating costs are lower and the cost of expensive fuel oil is avoided. Higher wind speed and solar radiation occur during summer, when the water demand is higher, and thus renewable systems can have lower installed power.

Renewable energy has been identified as one possible energy source for supporting desalination systems in general and in the Greek islands especially due to their high renewable energy potential in the forms of solar, wind, and in certain cases geothermal energy. Mean wind speed varies from 5 to 7 m/s, but in some areas it can be much higher. Annual solar radiation ranges from 1400 to 1700 kWh/m^2, and there are important high-enthalpy geothermal fields on some islands, such as Milos, Kimolos, Nisiros and Thira. Since the introduction of solar stills, numerous attempts have been made to employ renewable energy for desalination, both on pilot and commercial scales. The Greek state has promoted renewable-energy-powered desalination systems by giving priority in the licensing procedure to renewable systems and by providing a higher tariff for renewable energy sold into the grid.

Pilot units

A desalination unit of 80 m^3/day with a geothermal energy source, using multi-effect distillation technology, was constructed in the framework of a European project and operated successfully for demonstrations in Kimolos in 1997–98 (Karytsas, Alexandrou, & Boukis, 2002). The cost of water production was estimated at €1.7/m^3. At the end of the project, however, the desalination equipment was abandoned. A very small stand-alone pilot unit with a capacity of 4.8 m^3/day, powered by a 15 kW wind turbine, was installed in Therasia in 1997 and was in operation for many years. In 1998 a 500 kW wind turbine was installed on the island of Syros to drive a 900 m^3/day RO unit in stand-alone and grid-connected modes. The coupling of the two systems, however, was not achieved.

Hydriada, a floating RO desalination unit (80 m^3/day) with wind turbine and photovoltaics attached, was completed in 2007. It was a prototype designed to meet the water needs of 300 inhabitants on a small, arid island. The island of Iraklia was selected for the pilot application. Early technical problems with the unit, in addition to local-authority indifference and high maintenance costs, resulted in the project's being abandoned (Tsamopoulos, 2014).

Commercial units

A successful desalination plant that exploits wind energy was constructed on the island of Milos during 2007–09. The project consists of three similar RO units with a combined capacity of 3360 m^3/day and low energy consumption (approx. 3.5 kWh/m^3), powered by an 850 kW wind turbine. The RO units and the wind turbine are assisted by a supervisory control and data acquisition system to optimize the operation of the system as both the wind turbine and the desalination power load are connected into the island's autonomous grid. This optimization is achieved by the use of operation scenarios depending on weather and water demand forecasting and the current status of the water tanks. Another innovation of this project is its ownership status: the desalination unit and the wind turbine are private investments, subsidized by the state, and they work under contract partnership with the local municipality for providing water.

A mechanical vapour compression desalination system using wind energy (330 kW) was constructed in the island of Symi in 2009. This system was not successful due to technical problems that resulted in very low water production. The system was stand-alone, and its specific energy consumption was about 14.5 kWh/m^3 (Bundschuh & Hoinkis, 2012). Operation was halted in 2011 after a fire, and it was never repaired. On the very small island of Strogilli an autonomous RO unit of 8 m^3/day was installed in 2013 with the use of 20-kilowatt-peak photovoltaics. The unit covers the water needs of a small army camp.

Discussion and conclusions

Desalination has been adopted and used extensively as a solution for increasing water supply on the water-scarce Greek islands. Some initial efforts in the 1960s and 1970s using solar distillation plants were unsuccessful, mainly due to limited technical know-how and expertise, insufficient maintenance, and high operation and maintenance costs. Advances in desalination technology, together with the pressing need to meet demand in the rapidly developing tourist destinations, increased the prospects of desalination as a supplementary source of potable water. Both at the commercial and pilot levels, efforts have been made to enhance the performance of desalination. Technical limitations and economic considerations are the main reasons that such efforts were not always successful.

The main technical challenge is the reduction of the energy consumption of desalination units. As energy is a high cost factor for desalination, methods for lowering the specific energy consumption can help reduce costs. Another option is the use of cheaper energy sources. Renewable energy sources can offer a viable way to reduce pressure on energy systems, and thus developing renewable-energy-powered desalination plants is a priority in the Greek energy and water sectors, and a number of incentives are being considered. Energy storage is not a practical solution due to its high investment cost and the low lifetime of battery systems. Hybrid systems, combined with operation optimization schemes, can offer the appropriate solutions; however, wind parks and photovoltaics often face social opposition, especially in the tourism-dependent Greek islands, as locals find that the aesthetic value of the landscape is reduced. Environmental effects and land-use conflicts may also hinder the use of integrated renewable-energy-powered RO systems. Studies and pilot systems should be promoted that will lead to the optimal design and operation of renewable-energy-powered RO desalination systems. Historical data on daily water desalination production patterns and data on renewable energy sources can be used towards improved exploitation of renewable energy for water production.

Water deficiency in many arid and semi-arid islands is addressed with water transfers at the expense of the central government, whereas the operation and maintenance cost of desalination plants is subsidized, as, for social reasons, the true cost of water cannot be fully transferred to the locals (water users). In an era of economic crisis, this situation may not be viable, and there is a need to move from the established to more sustainable solutions. Different forms of private-sector involvement (as e.g. in the case of Milos), though not always socially acceptable, should be examined. In addition, as water and energy resources face increasing pressure over the next few decades, evaluating the trade-offs and encouraging cross-sector planning is crucial for their sustainable management and for development.

The Greek experience shows that building and operating desalination units is made possible by state support (in the form of subsidies). As seen from the comparative analysis presented in this article, the energy intensity of desalination plants remains high. However, RO technology, though costly in energy and financial terms, is reliable and provides adequate quality and quantity of potable water. Therefore, specific technical and policy challenges (including securing adequate funding) should be addressed so that RO technologies can continue to improve energy-efficiency and be part of an integrated water management system on the Greek Islands.

Overall, water supply is a politically sensitive topic on the islands. As a result of the lack of public awareness on water use and the technical problems of the early desalination units, a significant part of the population has formed a negative opinion of desalination systems. The state needs to inform and educate local populations on the long-term benefits of water desalination, while desalination plant owners have to maintain appropriate standards to ensure their efficient operation, taking the local environment into consideration. If social acceptance is achieved, especially in the tourism-dependent Greek islands, where the aesthetic value of the landscape is a crucial element of the local economy, renewable energy (wind parks and photovoltaics) could help in addressing energy challenges. An additional factor that should be addressed is the condition of water networks – poor on most islands and totally absent in many areas – which represents an obstacle for the widespread use of desalination. A flexible, independent public department or agency for the supervision of all desalination units for public supply could be created, in order to secure the application of good practices in all stages of water production, to test the final water quality, and to provide incentives for lowering energy demands. The "success stories" of desalination can be used as examples to promote it.

Water and energy security will be among the policy priorities in the next decades for Greece. Detailed assessments of the energy-for-water nexus for desalination can illustrate the differences between the various systems used in the islands and indicate options (best operational/management practices) for their improvement. Technology innovation will certainly increase energy efficiency; however, inter-regional and international cooperation will be necessary, as the challenge presented by the water-energy nexus is too large for any country, region, development finance institution or implementing agency to tackle alone.

Disclosure statement

No potential conflict of interest was reported by the authors.

References

Arnold, J. (1979). Operating experiences of a 15,000 cubic meter per day municipal desalting plant at Corfu, Greece. *Desalination, 30*, 9. doi:10.1016/S0011-9164(00)88442-8

Bundschuh, J., & Hoinkis, J. (2012). *Renewable energy applications for freshwater production*. Leiden: CRC Press/Balkema.

Cuenca, J. C. (2012). *Report on water desalination status in the Mediterranean Countries*. Murcia: IMIDA.

Dafnis, I. (2005). *Local reverse osmosis units of Corfu munincipality*. Potable Water: Quality Control, Legislation, Problems of Quality and Water Management in Corfu Prefecture (Ed.), Corfu: Technical Chamber of Greece.

Dagkalidis, A. (2009). *Greek sectorial studies - study on desalination*. Retrieved from http://www.piraeusbankgroup.com/~/media/Com/Downloads/Greek-Sectoral-Studies/2010/afalatosi_nerou.pdf

Dalmira, A. (2013). *Desalination unit in Schoinousa*. Retrieved from http://www.koinignomi.gr/news/politiki/politiki-kyklades/2013/12/27/monada-afalatosis-gia-ti-shoinoysa.html

Deligiannis, E. -E., & Belesiotis, V. (1995). Solar application in desalination: The Greek Islands experiment. *Desalination, 100*, 8. doi:10.1016/0011-9164(96)00006-9

EIP-Water. (2015). Water-energy nexus. Retrieved from http://www.eip-water.eu/priorities/water-energy-nexus

EL.STAT. (2012). *Announcement of the results of the 2011 population census for the resident population*. Retrieved from www.statistics.gr/portal/page/portal/ESYE/BUCKET/General/A1602_SAM01_DT_DC_00_2011_02_F_EN.pdf

EYATH, Thessaloniki Water Supply and Sewerage Company. (2010). *Annual bulletin*. Retrieved from http://www.eyath.gr/misc/ETISIO_DELTIO_2009.zip

EYDAP, Athens Water Supply and Sewerage Company. (2011). *Annual report & annual bulletin.* Retrieved from http://www.eydap.gr/userfiles/Presentations/ependutes/enimerotika-deltia/etisio_deltio_2011.pdf

Georgalas, L. (2010). *Approach of water-energy relation in the water supply-sewage system.* Retrieved from http://www.nomosphysis.org.gr/articles.php?artid=4237&lang=1&catpid=2

Hellenic Electricity Distribution Network Operator S.A. (2012). *Informative report of energy production on non-interconnected islands (June 2012).* Retrieved from http://www.deddie.gr/en/miniaia-deltia-ape-kai-thermikis-paragwgis-sta-mi-diasundedemena-nisia/2012

Hellenic National Meteorological Service (2015). Retrieved from http://www.hnms.gr/hnms/english/climatology/climatology_html

International Energy Agency (2012). *World energy outlook 2012.* Retrieved from http://www.iea.org/publications/freepublications/publication/WEO2012_free.pdf

imerisia.gr. (2009). New desalination unit from Temak. Retrieved from http://www.imerisia.gr/article.asp?catid=26519&subid=2&pubid=7421123&tag=9464#

IPCC (2014). *Climate change 2014: Impacts, adaptation, and vulnerability. Part A: Global and sectoral aspects. Contribution of working Group II to the fifth assessment report of the intergovernmental panel on climate change* (Report No. 978-1-107-05807-1). United Kingdom and New York, NY, USA: C. U. Press.

Kampourakis, N. (2013). *Deyath project report 2011–2012.* Retrieved from http://www.thira.gr/nomika-proswpa/deya-thiras/16-dimos/nomika-proswpa-synthesi/deya-thiras.html?layout=blog

Karagiannis, I. C., & Soldatos, P. G. (2007). Current status of water desalination in the Aegean Islands. *Desalination, 203*, 6. doi:10.1016/j.desal.2006.04.006

Karytsas, C., Alexandrou, V., & Boukis, I. (2002). *The Kimolos Geothermal desalination project.* Retrieved from https://pangea.stanford.edu/ERE/db/IGAstandard/record_detail.php?id=5365

Marsh, D. M. (2008). *The water-energy nexus: A comprehensive analysis in the context of New South Wales.* (Doctor of Philosophy (Engineering)), University of Technology Sydney, Sydney.

Nokas, H. (2002). *Estimation and comperative analysis of water production cost in the Cyclades Islands.* Syros Island: Water Directorate, Prefecture of Cyclades.

Rodriguez, D. J., Delgado, A., DeLaquil, P., & Sohns, A. (2013). *Thirsty energy.* Washington, DC: World Bank.

Schumacher, L. (2007). *Water - Energy nexus [powerpoint document].* Retrieved from http://www.naruc.org/Committees/CommitteePresentations.cfm?c=10

Scott, C. A., Pierce, A. S., Pasqualtti, J. M., Jones, L. A., Montz, E. B., & Hoover, H. J. (2011). Policy and institutional dimensions of the water–energy nexus. *Energy Policy, 39*(10), 6622–6630. doi:10.1016/j.enpol.2011.08.013

Special Secretariat for Water (2013-14). Integrated river basin management plans. Retrieved from http://wfd.ypeka.gr/.

sychem.gr. (2011). *Desalination of seawater in Paros Island.* Retrieved from http://www.sychem.gr/el/erga/afalatosi-yperdiithisi-erga/item/112-afalatwsh-thalassinoy-neroy-ston-dhmo-paroy

Tsamopoulos, M. (2014, August 7). *Cyclades: Desalination system is rotten in Iraklia.* Retrieved from http://www.protothema.gr/greece/article/400961/kuklades-katarrei-sustima-afalatosis-stin-irakleia-/

Vakondios, G. (2015). Potable water production with desalination. Paper presented at workshop on New Trends in Desalination Technology. Retrieved from http://www.edeya.gr/2013-09-23-10-58-06/2013-09-23-11-11-03/uliko/hmerides-1/texnologies-afalatwsis-anaduomenes-taseis-1

Ventouris, E. (2013). *Kimolos desalination plant.* Retrieved from http://www.gokimolos.gr/el/component/content/article/15-news/68-dimos-kimolou.html

Water Reuse Association (2011). *Seawater desalination power consumption.* Water Reuse Association - Desalination mmittee. Retrieved from http://www.watereuse.org/sites/default/files/u8/Power_consumption_white_paper.pdf

Modernization of irrigation systems in Spain: review and analysis for decision making

Javier Alarcón[a], Alberto Garrido[b] and Luis Juana[c]

[a]ETS Ingenieros Agrónomos, Universidad Politécnica de Madrid, Spain; [b]Departamento de Economía Agraria y Ciencias Sociales, ETS Ingenieros Agrónomos Research Centre for the Management of Agricultural and Environmental Risks (CEIGRAM), Universidad Politécnica deMadrid, Spain; [c]Departamento de Ingeniería Rural, ETS Ingenieros Agrónomos, Universidad Politécnica de Madrid, Spain

ABSTRACT

This article presents a method for analyzing the economic feasibility of modernizing irrigation systems. By using substitution relationships between two variables, one can determine irrigation performance and farm profit above which modernization would be justified from an economic point of view; and also the investments and energy consumption up to which modernization could be considered cost-effective. By means of representative average values, this method is applied to the now widespread conversion from surface irrigation to drip irrigation in Spain. Two conclusions are drawn. First, modernization may, in some instances, be justified only if more productive crop patterns are implemented. Second, saving water is a cheaper option than using an alternative resource only under certain conditions.

Introduction

For more than a decade, Spain's government has used a number of strategies to alleviate water shortages. One focus is on the modernization of irrigation, driven by the National Irrigation Plan (MAGRAMA, 2012) and the Emergency Plan for Modernization of Irrigation (BOE, 2006). Modernization covers actions aimed at improving transport networks in irrigated areas where water is currently used inefficiently. The policy program included repairing damaged and replacing obsolete irrigation networks, as well as rationalizing the layout of networks and changing irrigation systems. Modernization was justified by water savings, although other benefits were expected as well, such as higher irrigation productivity and competitiveness (Plusquellec, 2009).

With the modernization of irrigation systems over the last 10 years in Spain (2004–2014), the area under pressurized irrigation has increased by approximately 600,000 ha (90% with drip irrigation), while surface irrigation has been reduced by around 250,000 ha (MAGRAMA, 2015). This means that the total irrigated area has increased by about 350,000 ha, mostly in districts where more water resources are available. Over the past decade, the modernization programs in Spain have improved 2 Mha of irrigation and cost roughly €7400 M (Dumont, Mayor, & López-Gunn, 2013; MAGRAMA, 2012). The high investment costs and energy demand of pressurized irrigation have placed a huge burden on some farms with small margins, which account for a significant part of the Spanish farming industry. For this reason, in Spain as in other countries pursuing similar strategies, like Austria, Portugal, Netherlands, Australia and Mexico, many of the investments have been subsidized (Garrido & Calatrava, 2009; OECD, 2008). The pressure on water resources, coupled with an inadequate assessment of savings and costs, has unquestionably resulted in modernization projects becoming heavy sunk costs for the public administration or individual farmers (Hardy, Garrido, & Juana, 2012).

There is little evidence and few studies linking water use efficiency to real savings (Ward & Pulido-Velázquez, 2008). They include Huffaker and Whittlesey (2003), Peterson and Ding (2005), Schierling, Young, and Cardon (2006), and Hussain, Turral, Molden, and Ahmad (2007). When water efficiency improves, and thus the net endowment of water increases, the intended water savings are often compromised because some crops are intensified or irrigated areas are expanded, leading to less land being set aside during the summer. As a result, evapotranspirated water, which is irretrievably lost, increases. Additionally, while productivity per volume served at source may increase, the productivity per volume consumed is often the same or lower. This decreases irrigation returns (Lecina, Isidoro, Playán, & Aragües, 2010a, 2010b; Perry, Steduto, Allen, & Burt, 2009; Playán & Mateos, 2006; Rodríguez Díaz, Pérez, Camacho, & Montesinos, 2011b; Pfeiffer & Lin, 2012). The positive relationship between water efficiency and water use is often referred to as the 'hydrological paradox' (Ahmad, Giordano, Turral, Masih, & Masood, 2007; Jensen, 2007; Ward & Pulido-Velázquez, 2008; Rodríguez Díaz et al., 2011b) or the 'rebound effect' typical of the Jevons paradox (Alcott, 2005, 2008; Berbel, Gutiérrez-Martín, Rodríguez-Díaz, Camacho, & Montesinos, 2015; Tirado, Gómez, & Lozano, 2006).

This article examines the costs and benefits of the modernization of irrigation systems (i.e. the conversion from surface irrigation to drip irrigation) in Spain, using a simplified numerical analysis. Although subject to some assumptions and limiting conditions, simple modelling and average data, the results are relevant and may be helpful for making investment decisions and designing irrigation modernization programs.

Methods

Average data for modernization in Spain

Costs of modernization

All cost figures itemized in this article are stated in euros with respect to the year 2009. Costs are expressed per unit of irrigated area. Nevertheless, there are design elements that have a large impact on costs, which is not proportional to the modernized surface area (materials used, height of pumping, degree of automation, etc.). Investment costs relate to the cost of construction as per contract, not including VAT, or the costs of their implementation plus an additional 22% for overhead and business profit.

Investment in irrigation networks (distribution and delivery) was taken from CEDEX (2013), which analyzes 64 construction projects in Spain, of which 35 are infrastructure projects appraised here. The investment costs are sourced from these project budgets. Most include the conversion from open networks to buried piping, building of ponds, pumping stations and filtering system installation, remote automation and electrification. The developers of these construction projects were the Spanish State companies for agricultural infrastructure development and State companies for water management, as well as the river basin agencies. Geographically, 50% of the projects are located in the north-east quadrant of Spain and the Ebro River basin; 19% in the north of Spain and the Duero basin; 15.5% in the Mediterranean regions of Valencia and Murcia; 12.5% in central Spain or the Tagus basin; and the remaining 3% in the Guadalquivir basin. Significant cost differences between areas of Spain are not to be expected.

Investments in irrigation systems within farm boundaries or plots of land were taken from several sources, including López-Cortijo (2008), CHG (2010), and Camacho (2012). In the absence of specific projects, investments in the alteration of surface irrigation networks were taken from Playán, Slatni, Castillo, and Faci (2000) and Díaz Barcos (2001). Annual costs include operating costs, as well as the costs of upkeep and maintenance, including repair, replacement and cleaning. These costs were estimated based on Montero and Tarjuelo (2004), Corominas (2010), CHG (2010), Rodríguez Díaz, Camacho, and Blanco (2011a), and Camacho (2012).

The resulting average values for investment by irrigation area are given in Table 1. Figures in parentheses denote the minimum and maximum values, which are indicative of the wide range in costs. Figure 1 shows the distribution function of investment costs of conversion to drip irrigation.

Water use efficiency

Water use efficiency or performance is estimated as the fraction of water that meets the transpiration requirements of plants. It is therefore a complementary value to water loss. In irrigation, a distinction can be made between *water conveyance and distribution efficiency*, expressed as the ratio of the volume delivered at plots of land to the volume available at the headwaters of the irrigation network, and *application efficiency*, which refers to the ratio of water used to water supplied to the plots.

The documentation of modernization projects seldom provides estimates of water savings. A widespread and rough estimate is that they realize savings of 20–30%. Table 2 shows efficiency ranges for the three irrigation methods. These are the figures for networks that are in good condition, well designed and properly managed. Savings would be less under other conditions, and they should not be taken as reliable estimates for all projects. In the absence of better data, however, they are used as reference values.

Table 1. Average costs and cost ranges for the modernization of irrigation networks in Spain.

Type of modernization	Investment (€/ha)	Annual cost (€/ha)[1]
Replacing surface irrigation by drip irrigation	8,993 (5,922–14,035)	286 (202–370)
Modernization of distribution and delivery networks	6,039 (4,214–9,835)	
Changing application system: drip	2,954 (1,708–4,200)	
Replacing surface irrigation by sprinkler irrigation	7,916 (5,028–13,507)[2]	298 (204–393)
Modernization of distribution and delivery networks	6,039 (4,214–9,835)	
Changing application system: sprinklers	3,210 (2,500–3,920)	
Changing application system: pivots	1,542 (800–2,285)	
Improvement of surface irrigation	4,637 (3,591–6,815)[3]	109 (83–135)

[1]Annual cost of operation and upkeep/maintenance.
[2]Average costs in relation to the respective irrigation area where application systems have been changed both to sprinklers and to pivots.
[3]Including land replotting, levelling, topsoil removal, and construction of ditches, roads and perimeter drains.
Source: own elaboration from consulted references.

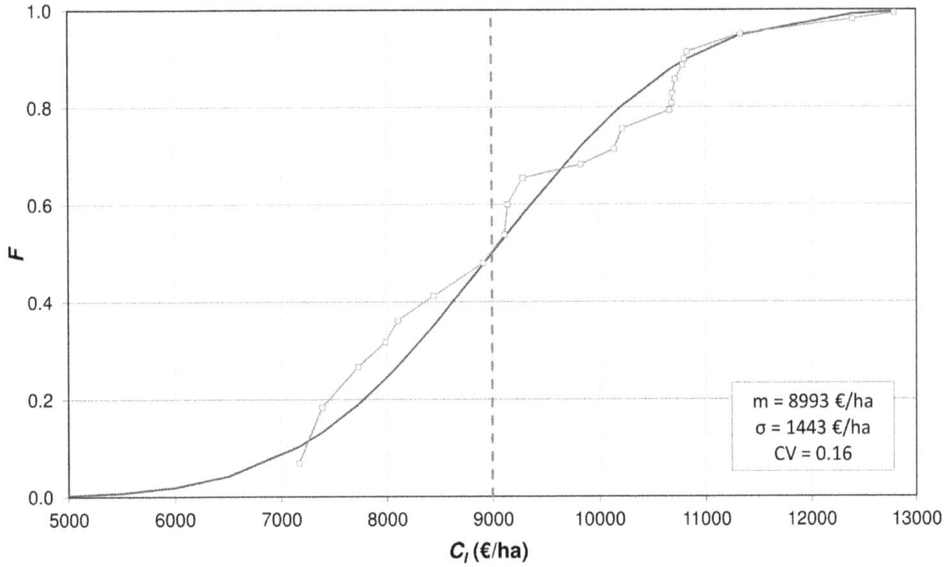

Figure 1. Probability distribution function of investments costs of transforming surface irrigation into drip irrigation in Spain. The solid line represents its fit to a cumulative normal distribution function. The dashed vertical line indicates the average investment. Frequencies (F) in absolute values. Note: m = average investment; σ = standard deviation; CV = coefficient of variation. Source: Author's own elaboration from CEDEX (2013).

Performance indicators and 'benchmarking' have been used to evaluate irrigation efficiency (Alexander & Potter, 2004; Lecina et al., 2010b; Mateos et al., 2010). These tools have a lot of potential for determining whether an action can achieve a particular objective. Previous studies have compared irrigated areas based on the productivity of their water (Allan, 1999; Rodríguez Ferrero, Salas, & Sánchez, 2010; Rodríguez Díaz, Camacho, López, & Pérez, 2008, 2012). In addition, the cost of the water saved has been used to study the effectiveness of certain measures (Semaan, Flichman, Scardigno, & Steduto, 2007; Turner, Georgiou, Clark, Brower, & Burke, 2004; Zanou, Kontogianni, & Skourtos, 2003; Blanco-Gutiérrez, Varela-Ortega, & Flichman, 2011). In this article, four indicators are considered useful for characterizing the modernization of the analyzed irrigation networks: (1) water savings related to investment and energy consumption; (2) the benefit that offsets the investment, either by increasing operating margins or the use made of water saved; (2) the cost of each cubic metre saved; and (4) water productivity.

Procedures

An irrigation modernization project leads to changes in the use of water, energy and capital. Even if the production orientation is unchanged, the availability of new facilities wil enable changes in the production process, which may affect profits. Given that some reference parameters are uncertain and ever-changing, it is useful to establish relationships and intervals in order to single out when the values of those profits indicate that modernization projects turn out to be profitable. This approach is applicable for both public and private purposes. If interpreted broadly, the following variables can account for very different, and even indirect, costs and benefits for either public administrations or individual farmers.

Table 2. Ranges of common efficiency ratios in conveyance (Rr) and application networks (Ra) for different irrigation systems; average and standard deviation efficiencies in joint networks (R).

			$R = R_r \cdot R_a$		
Irrigation system	R_r	R_a	Mean	Std. dev.	CV
Surface	0.65–0.95	0.50–0.80	0.54	0.20	0.37
Sprinklers	0.75–0.95	0.70–0.90	0.70	0.15	0.22
Drip	0.75–0.95	0.90–0.95	0.79	0.10	0.13

Note: CV: coefficient of variation (= std. dev. / mean).
Source: own elaboration from Krinner, García, and Estrada (1994), Hanson (1995), Rogers et al. (1997), SWSI (2004), and ESTEPA (2010).

When comparing two situations, before and after modernization, the variation in the water use or irrigation performance efficiency, $R_2 - R_1$, and the variation in operating profit, B (€/ha per year), are expected to be positive, because otherwise the change would certainly not be of interest. But of course there will also be increases in costs: (1) investment, C_I (€/ha); (2) the annual cost of operation and maintenance of infrastructure or facilities and the attributable administrative expenses, C_M (€/ha per year); (3) installed power, P (kW/ha); (4) energy consumption, E_v (kWh/m^3); and (5) cost of water, C_w (€/ha per year), subject to the inclusion of fees with respect to the recovery of costs associated with the investment. For simplicity's sake, the article includes the cost of water in C_I and C_M. The sign of the variation in water requirements, ET (m^3/ha per year) and water savings, S (m^3/ha per year), will be different across projects, although it is reasonable to assume that the sign for this last parameter will be positive, as water demand should plausibly drop ($\Delta W < 0$).

$$\begin{array}{lll} \Delta B = B_2 - B_1 & S = -\Delta W = \frac{ET_1}{R_1} - \frac{ET_2}{R_2} & \Delta C_I = C_{I2} - C_{I1} \\ \Delta C_M = C_{M2} - C_{M1} & \Delta P = P_2 - P_1 & \Delta E = \frac{ET_2}{R_2} \cdot E_{v2} - \frac{ET_1}{R_1} \cdot E_{v1} \end{array} \tag{1}$$

To check whether a final Scenario 2 is preferable to an initial Scenario 1, the increased profit achieved should be proven to outweigh the increase in costs. Water savings can be valued by assigning the marginal benefit that can be earned for a particular use, MB (€/m^3), or alternatively the marginal cost, MC (€/m^3), of earning the benefit from an alternative source. Consequently, the modernization project will be justified if the following inequality holds:

$$S \cdot MB + \Delta B - \Delta C_I \cdot a - \Delta C_M - \Delta P \cdot c_{Pu} - \Delta E \cdot c_{Eu} \geq 0 \tag{2}$$

where a (y^{-1}) is the parameter annualizing the investment; (2) c_{Pu} (€/kW per year) is the fixed costs of electrical power (note that C_I includes the investment costs of the pumping plant); and (3) c_{Eu} (€/kWh) is the price of energy. For simplicity's sake, operating/maintenance and electrical power costs have been discounted from the operating profit in Equation (2s), because they can both be considered fixed costs: $B' = B - C_M - P \cdot c_{Pu}$. Although this figure could continue to be called the 'operating profit', it is differentiated by adding a prime mark (′) to avoid confusion.

$$S \cdot MB + \Delta B' - \Delta C_I \cdot a - \Delta E \cdot c_{Eu} \geq 0 \tag{2s}$$

Unit water savings

Combining $\Delta B'$ with Equation (1) above for ΔS and ΔE yields:

$$\frac{ET_1}{ET_2} \cdot \frac{R_2}{R_1} - 1 \geq \frac{\Delta C_I \cdot a - \Delta B'}{MB} \cdot \frac{R_2}{ET_2} + \left(1 - \frac{ET_1}{ET_2} \cdot \frac{R_2}{R_1} \cdot \frac{E_{v1}}{E_{v2}}\right) \cdot \frac{E_{v2} \cdot c_{Eu}}{MB} \tag{3}$$

Provided that all the variables in the inequality (Eq. 3) but one (e.g. R_2 or MB) are known, the point of equalization can be used to find the value of the unknown variable as of which the second situation would begin to be of interest. Thus, it may well be useful to fix values for the known variables in order to determine this critical value of the unknown variable. The initial scenario of the conversion from surface irrigation to pressurized irrigation is assumed not to require energy ($E_{v1} = 0$). In order to simplify Equation (3) and identify or isolate some effects of modernization, e.g. water savings, the net water needs are also assumed to be equal ($ET_1 = ET_2 = ET$). In spite of this, the abstraction requirements differ, because each technique has a different irrigation performance ($R_1 \neq R_2$). Note, however, that this methodology could also handle different *ET*s.

Given the above assumptions, Equation (3) can then be written as:

$$\frac{R_2}{R_1} - 1 \geq \frac{\Delta C_I \cdot a - \Delta B'}{MB} \cdot \frac{R_2}{ET} + \frac{E_{v2} \cdot c_{Eu}}{MB} \tag{4}$$

The left-hand side of Equation (4) refers to unit water savings with respect to the initial scenario. Note that it may take values greater than 1. The first term on the right-hand side represents the variation of costs and benefits, mainly due to investments and operating margins; and the second term, the variation in energy expenditure. Both are expressed with respect to the marginal benefit of water saved. We find that if $\Delta B' = \Delta C_I \cdot a$, the change would be justified if water savings are greater than the equivalent energy consumption. When this change does not imply energy consumption ($E_{v2} = 0$), the project would be justified if water savings are greater than the equivalent variation of costs and benefits.

Benefit and marginal cost of water saved

If operating profit is unchanged, $\Delta B = 0$, modernization would only be justified by potential water savings, *S*. If the marginal benefit achieved with the water saved, *MB*, is greater than its marginal cost, the investment is of interest. Then, for the above case where $ET_1 = ET_2 = ET$, the result of finding *MB* in Equation (2) and writing ΔS likewise in Equation (1) is:

$$MB = (\Delta C_I \cdot a + \Delta C_M + \Delta P \cdot c_{Pu} + \Delta E \cdot c_{Eu}) \cdot \frac{R_1 \cdot R_2}{ET(R_2 - R_1)} \equiv MC \tag{5}$$

On the other hand, the marginal cost of water saved, *MC*, is defined as the ratio between the total cost of the modernization action and the water saved, ΔS. Analytically, this ratio is the same as Equation (5), which is also valid for this indicator. *MC* is potentially useful insofar as it can be compared with the costs of procuring new water resources, instead of saving water.

Water productivity

A distinction should be made between the marginal benefit that can be gained from the water saved, *MB*, and the production value of water to be achieved after modernization, P_{w2}. Final productivity is the result of subtracting the sum of investment and operating/maintenance costs from the operating profit, and then dividing by the volume of water that is supplied to the irrigation network. Considering the simplification made in Equation (2s):

$$P_{w2} = (B' - C_I \cdot a - E_2 \cdot c_{Eu2}) \cdot \frac{R_2}{ET_2} \tag{6}$$

Application to a case study with reference values for Spain

The above methods were applied to the conversion of surface irrigation to drip irrigation. The average investment shown in Table 1, C_{I2} = €8993/ha, and the average irrigation performance in Table 2, $R_2 = 0.79$, were assumed to be representative at the national level. The upkeep/maintenance costs were estimated at 1.2% of the investment, according to MMA (2000). Energy costs were estimated as: $P_2 = 1.5$ kW·h/ha (Rodríguez Díaz et al., 2011b); C_{P2} = €44.28 kW/y (Ederra, 2010); $E_{v2} = 0.53$ kW·h/m^3 (Corominas, 2010), assuming $E_{v1} = 0$; c_{Eu} = €0.074/kWh (Ederra, 2010).

The investment lifetime was assumed to be 25 years, with a depreciation rate and interest rate of 4%, which is a typical rating for such projects. Hence, the annuity used is $a = 0.064$/y. Given that some authors, like Scherer and Weigel (1993), have recommended a 15–20 year project lifetime for pressurized systems, results were also calculated for a 20-year project lifetime. The proportion between the respective annuities, $a_{20}/a_{25} = 1.15$, gives a good idea of the required ratio of all the discussed costs to benefits.

Annual water needs were assumed to be $ET = 4116$ m^3/ha, which is consistent with the data on average net demand for Spain provided by the National Irrigation Plan (MAGRAMA, 2012). Given that the aim of the study is to assess the economic effect of the modernization of irrigation works, we assume that $ET_1 = ET_2 = ET$ in this case study. Note, however, that this does not necessarily imply that the crop pattern has to remain unchanged. In practice, crop patterns are often changed whereby, after modernization, either more profitable crops are introduced or a larger surface area is devoted to such crops or even to irrigation (as envisaged in MARM, 2010).

To complete the method, the variation in the operating margin due to modernization must be estimated. To do this, data were used from the Spanish Genil MI irrigation area in the province of Córdoba, Andalusia. These data were sourced from the feasibility report on the respective modernization project (MARM, 2010), as well as other technical and economic farm data for Andalusia (Junta de Andalucía, 2002; MAGRAMA, 2010). We estimated specifically for Genil MI that, subsequent to irrigation modernization, the resulting improvement in water use increased production as follows: 15% for corn, 19% for cotton, 20% for oranges and 14% for outdoor vegetables (MAGRAMA, 2010; MARM, 2010).

The conversion from surface irrigation to drip irrigation in Genil MI led to a noticeable increase in orange trees, mainly at the expense of cotton and maize. As a result, however, water needs hardly changed at all and were close to the average *ET* given above. Results for this example are reported considering crop patterns both before and after the conversion.

Profit versus investment: benefits that outweigh the modernization

When attempts to save water fail ($R_2 = R_1$) or no benefit is attained from the water saved (MB = 0), the increased operating profit required to compensate the investment, as well as the new energy expenditure, can be calculated. This equation is based on Equation (4).

$$\Delta B \geq \Delta C_I \cdot a + \frac{ET}{R_2} \cdot E_{v2} \cdot c_{Eu} \tag{7}$$

With the average values specified above, fixed costs are $\Delta B'$ = €780/ha, with $\Delta B' = \Delta B - \Delta C_M - \Delta P{\cdot}c_{Pu}$ (see section "Application to a Case Study"); the result is ΔB = €903/ha. At this value

Data	mean	CV
ΔC_I (€ ha^{-1})	8993	0.16
a	0.064	-
ET (m^3 ha^{-1})	4116	0.15
R_2	0.79	0.20
$E_{v2} \cdot c_{Eu}$ (€ m^{-3})	0.039	0.27
$\Delta P \cdot c_{Pu}$ (€ ha^{-1})	66	0.68
ΔC_C (€ ha^{-1})	57	0.72
$\Delta B_{estimated}$ (€ ha^{-1})	411	-
Results	**mean**	**CV**
$\Delta B_{required}$ (€ ha^{-1})	903	0.16

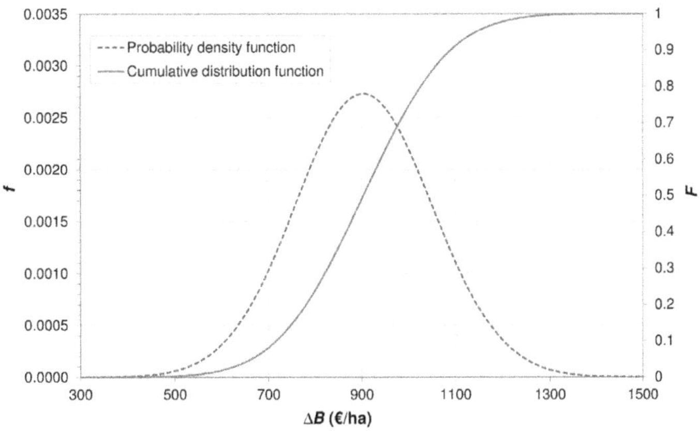

Figure 2. Probability density function (f) and cumulative distribution (F) of the increase in operating profit required for the conversion from surface irrigation to drip irrigation.

(required ΔB), the investment would be profitable. This value may be lower, considering the benefit of water saved. In the Genil MI irrigation area (Córdoba), under the assumption of maintaining the original cropping pattern after modernization, the average increase in the operating margin estimated for the crops amounts to only ΔB = €411/ha ($\Delta B'$ = €288/ha). Consequently, margins would be insufficient to justify the investment in modernization in this case. In fact, once the project was finished, a larger area was devoted to the more profitable crops (oranges and vegetables) at the expense of the others (corn and cotton). According to the data used, this could amount to an increment of €1380/ha in the operating margin there, high enough to capitalize investment.

With the mean values and coefficients of variation of the data specified above (Tables 1 and 2), we estimate the average value and the coefficient of variation (CV) for ΔB. To do this, we apply classical approximate relationships (Benjamin & Cornell, 1970) to Equation (7), and also to the expression that relates ΔB to $\Delta B'$ (from combining Equations (2) and (2s)). If ΔB is also assumed to follow a standard distribution, we can calculate its probability density function (Figure 2). According to Figure 2, the probability of, for example, ΔB = €411/ha outweighing the costs would be just 0.01%.

Water savings that justify modernization

Equation (4) can be used to consider the relative increase in performance or overall irrigation efficiency, $R_2/R_1 - 1$, which represents the 'unit water savings' with respect to the initial scenario. This indicator can be used to justify an increase in the energy demand of an investment.

Water savings versus energy consumption

According to Equation (4), if $\Delta C_I \cdot a = \Delta B'$ (Assumption 1 in Table 3), a unit water saving, $R_2/R_1 - 1$, justifies an increase in energy consumption given by:

$$E_{v2} = (R_2/R_1 - 1) \cdot MB/c_{Eu} \tag{8}$$

Table 3. Initial and final irrigation performances, R_1 and R_2, and unit water savings, $R_2/R_1 - 1$, for the conversion from surface irrigation to drip irrigation. Energy consumption and related investments under different assumptions.

	$\frac{R_2}{R_1} - 1$		0.05	0.10	0.15	0.20	0.25	0.30	0.40	0.50	0.60	0.75	1.00
Assumptions	F	R_1						R_2					
	0.22	0.30	0.32	0.33	0.35	0.36	0.38	0.39	0.42	0.45	0.48	0.53	0.60
	0.27	0.35	0.37	0.39	0.40	0.42	0.44	0.46	0.49	0.53	0.56	0.61	0.70
	0.32	0.40	0.42	0.44	0.46	0.48	0.50	0.52	0.56	0.60	0.64	0.70	0.80
	0.38	0.45	0.47	0.50	0.52	0.54	0.56	0.59	0.63	0.68	0.72	0.79	0.90
	0.45	0.50	0.53	0.55	0.58	0.60	0.63	0.65	0.70	0.75	0.80	0.88	1.00
	0.51	0.55	0.58	0.61	0.63	0.66	0.69	0.72	0.77	0.83	0.88	0.96	
	0.57	0.60	0.63	0.66	0.69	0.72	0.75	0.78	0.84	0.90	0.96		
	0.64	0.65	0.68	0.72	0.75	0.78	0.81	0.85	0.91	0.98			
	0.70	0.70	0.74	0.77	0.81	0.84	0.88	0.91	0.98				
	0.75	0.75	0.79	0.83	0.86	0.90	0.94	0.98					
	0.80	0.80	0.84	0.88	0.92	0.96	1.00						
	0.84	0.85	0.89	0.94	0.98								
	0.88	0.90	0.95	0.99									
1. $B'_u = a \cdot C_{Iu}, E_u \neq 0$	E_{v2} (kWh/m)		0.20	0.39	0.59	0.78	0.98	1.18	1.57	1.96	2.35	2.94	3.92
	H (m)		58	115	173	230	288	346	461	576	691	864	1,152
2. $E_u = 0, B'_u = 288, R_2 = 0.79$	C_{Iu} (€/ha)		1,468	2,648	3,829	5,009	6,189	7,369	9,730	12,090	14,450	17,991	23,892
	F		0.00	0.00	0.00	0.00	0.03	0.13	0.70	0.98	1.00	1.00	1.00
3. $E_u \neq 0, B'_u = 288, R_2 = 0.79$	C_{Iu} (€/ha)		2,488	3,668	4,848	6,028	7,209	8,389	10,749	13,110	15,470	19,011	24,912
	F		0.00	0.00	0.00	0.02	0.11	0.34	0.89	1.00	1.00	1.00	1.00

Notes: $\Delta B'$: increase in operating profit, excluding upkeep and maintenance costs and the fixed cost of energy, $\Delta B' = \Delta B - \Delta C_M - \Delta P \cdot c_{Pu}$.
E_{v2}: energy consumption per m^3 of water in the modernized scenario, where, in this case:
$E_{v1} = 0$; $E_{v2} = (R_2/R_1 - 1) \cdot MB/c_{Eu}$ with $MB = €0.29/m$ and $c_{Eu} = €0.074/kWh$.
H: pumping height for E_{v2} and a pumping performance of $\eta = 0.8$.
ΔC_I: investment cost found in expression (4):
$\frac{R_2}{R_1} - 1 \geq \frac{\Delta C_I \cdot a - \Delta B'}{MB} \cdot \frac{R_2}{ET} + \frac{E_{v2} \cdot c_{Eu}}{MB}$, varying according to the table, with
$a = 0.064/y$, $\Delta B' = €288/ha$, $MB = €0.29/m^3$, $R_2 = 0.79$, $ET = 4{,}116\ m^3/ha$, $E_{v2} \cdot c_{Eu} = €0.039/m^3$.
F: probability of the respective entry not being exceeded.

where *MB* is the marginal benefit that can be expected from water in the final scenario. E_{v2} will be given in kWh/m^3, or as a pumping height *H* (m), once a given performance of the pumping station, η, is considered.

Table 3 accounts for MB = €0.29/m^3 and an energy cost of c_{Eu} = €0.074/kWh. With these values, the result for a unit energy consumption of E_{v2} = 0.59/kW·h per m^3 (equivalent to an elevation of 173 m, with $\eta = 0.80$) is $R_2/R_1 - 1 = 0.15$. Thus, if it allowed for a final R_2 of at least 0.69, modernization would be justified for an initial irrigation performance of $R_1 = 0.60$. As shown in the Table 3, it was inferred from the literature review that in $F = 57\%$ of the irrigated area in Spain performances are lower. Note that *MB* has a decisive influence. Therefore, if *MB* were four times lower, the same value of E_v or *H* would be equivalent to an $R_2/R_1 - 1$ value four times higher (0.60). This value would make the modernization unprofitable for most farms, as R_2 would have to be at least 0.96. It is also possible to estimate the minimum *MB* below which conversion would be of no interest.

Water savings versus investment costs

Assumption 2 in Table 3 is equivalent to the case where energy was not increased, $\Delta E = 0$, and modernization would achieve reference values of $R_2 = 0.79$ and $\Delta B'$ = €288/ha per year. Solving for ΔC_I in Equation (4) yields:

$$\Delta C_I = \frac{1}{a} \cdot \left(\frac{R_2}{R_1} - 1\right) \cdot \frac{ET}{R_2} \cdot MB + \Delta B' \qquad (9)$$

Thus, a given water saving, $R_2/R_1 - 1$, could justify some investment, ΔC_I. By entering an investment cost in Table 3 (last but one row), we can calculate the initial irrigation performance, R_1, below which conversion would be of interest. Thus, an investment of €7369/ha, which would cover up to $F = 13\%$ of the irrigated area in Spain, would correspond to $R_2/R_1 - 1 = 0.30$. Since $R_2 = 0.79$, modernization would be justified if R_1 is less than 0.60 (57% of the irrigated area); but this investment could not be justifiable for a higher R_1.

Water savings versus investment costs and energy

Because there is investment and variation in energy consumption, the two addends on the right-hand side of Equation (4) must be considered. This scenario corresponds to Assumption 3 in Table 3. For ΔC_I = €8993/ha and the other reference variables considered in this case study (see Figure 2), the general entry in Table 3 would now be $R_2/R_1 - 1 = 0.32$. Thus, for a final irrigation performance of $R_2 = 0.79$, the initial performance should be below 0.60. Although inversion costs were up to €10,800/ha in the Genil MI irrigation area, modernization would have been justified from this point of view, too, because the initial performance was below 0.58, and especially since the change in the crop pattern resulted in an increment in the operating margin.

The impact of energy consumption, E_{v2}, and energy cost, ΔC_E, is noteworthy. A 50% increase in the energy cost used in Table 4, up to €0.111/kWh, would result in $R_2/R_1 - 1$ = 0.39; hence, for $R_2 = 0.79$, the initial irrigation performance, R_1, should not be smaller than 0.57. Moreover, if c_{Eu} = €0.148/kWh, that is, two times the reference energy cost, $R_2/R_1 - 1$ = 0.46 and $R_1 < 0.54$, which is precisely the average efficiency sourced from the literature review for surface irrigation in Spain. This means that, according to these data, water savings from the conversion to drip irrigation would not outweigh costs in about 50% of the cases for this type of irrigated land. A similar analysis can be conducted by fixing R_1 instead of R_2.

Table 4. Benefits and marginal costs of water saved, MB ≡ MC, for the conversion of surface irrigation to drip irrigation, with a final irrigation performance $R_2 = 0.79$, related to the water needs, ET, the unit water savings, $R_{2/}R_1 - 1$ and the initial performance, R^1.

ET (m³/ha)		1500	2100	2700	3300	3900	4500	5100	5700	6300	6900	7500
$\frac{R_2}{R_1} - 1$	R_1	MB ≡ MC (€/m³)										
1.63	0.30	**0.28**	**0.20**	**0.16**	**0.13**	**0.11**	**0.09**	**0.08**	**0.07**	**0.07**	**0.06**	**0.06**
1.26	0.35	0.37	**0.26**	**0.20**	**0.17**	**0.14**	**0.12**	**0.11**	**0.10**	**0.09**	**0.08**	**0.07**
0.98	0.40	0.47	0.34	**0.26**	**0.22**	**0.18**	**0.16**	**0.14**	**0.12**	**0.11**	**0.10**	**0.09**
0.76	0.45	0.61	0.44	0.34	**0.28**	**0.24**	**0.20**	**0.18**	**0.16**	**0.15**	**0.13**	**0.12**
0.58	0.50	0.80	0.57	0.44	0.36	0.31	**0.27**	**0.23**	**0.21**	**0.19**	**0.17**	**0.16**
0.44	0.55	1.06	0.76	0.59	0.48	0.41	0.35	0.31	**0.28**	**0.25**	**0.23**	**0.21**
0.32	0.60	1.46	1.04	0.81	0.66	0.56	0.49	0.43	0.38	0.35	0.32	**0.29**
0.22	0.65	2.15	1.53	1.19	0.98	0.83	0.72	0.63	0.56	0.51	0.47	0.43
0.13	0.70	3.59	2.57	2.00	1.63	1.38	1.20	1.06	0.95	0.86	0.78	0.72
0.05	0.75	8.66	6.19	4.81	3.94	3.33	2.89	2.55	2.28	2.06	1.88	1.73

Note: $R_2 = 0.79$, $ET_1 = ET_2 = ET$, $a = 0.064$/y (25 years) are considered.

Marginal benefit and cost of water saved: water productivity

Marginal benefit of water saved

The *MB* value of saved water estimated for the modernization of a certain irrigation area should be rated with respect to the reference *MB* value found for that particular area. In such a way, if the former does not exceed the reference value, the modernization will be of interest. Using Equation (5), and by changing R_1 and *ET*, we calculated the marginal benefits of water saved, *MB*, which would compensate the average costs considered for the conversion from surface irrigation to drip irrigation. These *MB* values are shown in Table 4. Although it is a different matter, reference water productivity has been used to exemplify this point. In Spain, the average value of water productivity given in MIMAM (2007) is €0.29/m³. A low productivity value that could also be used for comparison is €0.11/m³, which refers to the Duero River basin. The application of this method should, of course, adopt the best suited value in each case.

Thus, *MB* values not exceeding the average water productivity in Spain (€0.29/m³) are highlighted in bold font in Table 4. Their respective *ET*, $R_2/R_1 - 1$ and R_1 combinations are the ones that justify investment, because with such values it is possible to earn more profit than is strictly necessary to compensate investment. In line with the above, initial water needs are unchanged (assuming $ET_1 = ET_2$). As shown in Table 4, modernization is unlikely to be economically feasible for productivity values lower than €0.11/m³, which can be found in some irrigation areas, such as the Duero River basin in Spain. According to the productivity values reported by Alarcón (2011), modernization would have been justified with respect to the Genil MI irrigation area for an R_1 below 0.55 if area growing oranges, and fruit and vegetables were expanded at the expense of corn and cotton, as was the case. On the other hand, if the original crop pattern had been maintained, R_1 would have had to be below 0.25.

Marginal cost of water saved

It is conceivable that modernization would be justified if the marginal cost of the water saved, *MC* (Equation (5)), is less than the cost of procuring water by other means. As a baseline value, the article considers a cost for desalinated water of €0.50/m³, which has been approximated from Torres (2004) and various feasibility reports on desalination plants (ACUAMED, 2006a,

2006b; ACUASEGURA, 2007). This cost is almost unaffordable for most Spanish agricultural producers at present, although the average water productivity is larger in the Mediterranean Basin of Andalusia, the Canary Islands and the Segura basin (MIMAM, 2007). The cost of providing an alternative resource such as regenerated water or water transfer could be likewise considered.

Table 4 shows that roughly over half of the *MC* values do not exceed that desalination cost. Their respective *ET*, $R_2/R_1 - 1$ and R_1 combinations will be the ones that justify the investment. Thus, by comparison with that *MC* value, we find that generally, except for high water needs, drip irrigation would not be justified when the initial performance, R_1, is greater than 0.60. For $ET = 4116$ m³/ha and the other adopted reference values, such investment would be justified for an R_1 lower than 0.60.

Water productivity

For the modernization projects considered in the case study, Equation (6) yields a water productivity of €0.22/m³. Assuming that initial irrigation performance is $R_1 = 0.45$, and considering $ET = 4116$ m³/ha, the productivity would be €0.18/m³. In this case, the productive value of water will improve. The assumption that the original crop pattern is maintained reduces the productivity value compared with farms in modernized areas where high-yield crops have been introduced (as is the case with citrus fruits, lemon, mandarin, and oranges). Thus, according to Alarcón (2011), productivity could reach €0.38/m³ for our running example in the Genil MI area, with crop reallocation and new orange tree plantations accounting for 70% of the land. Berbel, Kolberg, Rodríguez Díaz, and Montesinos (2007) calculated €0.34/m³ for citrus fruits in the Guadalquivir Valley. Finally, according to Fernández García, Rodríguez Díaz, Montesinos, Camacho, and Berbel (2013), the modernization of four irrigated areas of Andalusia, along with the substitution of crops such as corn and cotton by citrus fruits, has increased productivity from €0.40–0.70/m³ to €0.66–1.09/m³.

Conclusions

Investments and annual costs of modernized irrigation systems in Spain were estimated using broad data sources and parameters, including mean values and ranges of variation. The article also estimated their water use efficiency. Although some specific scenarios may not be properly represented, the respective values can, largely speaking, be considered as reference values for the proposed study. They do not differ significantly from the values used by Spanish government agencies. More detailed data on specific regions will allow more accurate estimates.

The method used here is able to estimate the increases in private operating profit to justify investments in modernization projects. Based on average costs, it can be inferred that the conversion to drip irrigation in some Spanish irrigation systems is unlikely to be profitable without a reorientation of production that considerably increases farm margins. However, profitability would be more easily achieved if the benefit from the water saved is taken into account. Presumably, present and future modernizations will be subject to the viability of more profitable crops, as well as to consideration of the opportunity cost of water.

It has also been shown that modernization would result in water savings only under certain circumstances. This is because the cost attributable to water saved does not seem to be offset by its potential benefit. Thus, savings are likely to occur only if the irrigation performance of the traditional methods is low or the irrigation needs are high. Moreover, a greater water delivery efficiency may induce more land to be irrigated, resulting in a 'rebound effect' of higher water consumption. The cost of water saved may be even higher than an alternative resource, such as desalinated water. Cheaper alternatives, such as increasing the use of surface irrigation, should be considered. Additionally, it is important to ensure proper maintenance of existing facilities so that they continue to operate according to design specifications.

A more conclusive analysis of the costs and benefits of irrigation modernization would require detailed specific information. For example, a similar technical option for farmers may incur very different costs and generate variable outcomes, depending on specific characteristics. Furthermore, it is rather difficult to assess the economic benefits of the significant social advantages generated by modernization, which have not been considered. Therefore, rather than concluding that one solution is better than another, the aim of the study was to see which new performance levels or which new farm margins justified which conversions.

Likewise, the analysis proposed here should be rounded out with a study of water returns and the possibility of reusing unconsumed water. Also, the allocation of water rights should be revised based on the improvements in irrigation performance in order to reduce the non-productive consumption of water. Another aspect worth considering is the quality of water returns, as they may increase with irrigation modernization. Finally, the possible destruction of the hydrophilic vegetation and habitats that traditional irrigation systems usually harbour should be considered. This is because, as described above, some actions aimed at increasing irrigation efficiency may incur meagre savings of water at a high cost.

Future studies should include the above externalities, which were not addressed in the article but which could be added to the proposed formulation. Despite the difficulty of valuing the indirect and non-market benefits of water and modernization, the results would help decision makers decide whether to undertake a modernization initiative, as well as how to best manage water resources. The productive value of water and the costs of converting an unproductive into a productive unit of water (i.e. desalination) are necessary reference values for effectively estimating these costs and benefits.

Disclosure statement

No potential conflict of interest was reported by the authors.

References

ACUAMED. (2006a). Informe de viabilidad de la Actuación 2.1.e. Nueva desaladora de Águilas/Guadalentín. Ampliación de la desaladora de Águilas (planta desaladora para el riego en Murcia) [Feasibility report of Action 2.1.e. New desalination plant of Águilas/Guadalentín. Expansion of the desalination plant of Águilas (desalination plant for irrigation in Murcia)]. Aguas de las Cuencas Mediterráneas, S.A. (ACUAMED) [In Spanish].

ACUAMED. (2006b). Informe de viabilidad de la Actuación 2.1.B. Planta desaladora para garantizar los regadíos del trasvase Tajo – Segura [Feasibility report of Action 2.1.B. Desalination plant to ensure the irrigation area of the Tajo-Segura water transfer]. Aguas de las Cuencas Mediterráneas, S.A. (ACUAMED) [In Spanish].

ACUASEGURA. (2007). Desalinizadora de Valdelentisco, en Cartagena (Murcia). Informe de viabilidad [Desalination plant of Valdelentisco, in Cartagena (Murcia). Feasibillity report]. Aguas de las Cuencas del Segura (ACUASEGURA) [In Spanish].

Ahmad, M. D., Giordano, M., Turral, H., Masih, I., & Masood, Z. (2007). At what scale does water saving really save water? Lessons from the use of resource conservation technologies in Pakistan. *J Soil Water Conserv, 62*, 29A–35A.

Alarcón, J. (2011). Evaluación de la eficiencia de un proyecto típico de modernización de regadíos [Efficiency assessment of a typical project for the modernization of irrigation networks]. *Revista Española de Estudios Agrosociales y Pesqueros, 230*, 119–145. [In Spanish].

Alcott, B. (2005). Jevons paradox. *Ecological Economics, 54*, 9–21.

Alcott, B. (2008). Historical overview of the Jevons paradox in the literature. In J. M. Polimeni, K. Mayumi, & M. Giampietro (Eds.), *The Jevons paradox and the myth of resource efficiency improvements* (pp. 7–78). London: Earthscan.

Alexander, P., & Potter, M. (2004). Benchmarking of Australian irrigation water provider businesses. *Irrigation and Drainage, 53*, 165–173.

Allan, T. (1999). Productive efficiency and allocative efficiency: Why better water management may not solve the problem. *Agricultural Water Management, 40*, 71–75.

Benjamin, J. R., & Cornell, C. A. (1970). *Probability, statistics and decision for civil engineers*. New York, NY: McGraw-Hill Book Company.

Berbel, J., Gutiérrez-Martín, C., Rodríguez-Díaz, J. A., Camacho, E., & Montesinos, P. (2015). Literature review on rebound effect of water saving measures and analysis of a Spanish case study. *Water Resources Management, 29*, 663–678.

Berbel, J., Kolberg, S., Rodríguez Díaz, J. A. , & Montesinos, P. (2007). EU water saving potential II, final report (ENV.D.2/ETU/2007/0001r). *Ecologic - institute for international and European environmental policy* (pp. 2–33). Berlin: University of Kassel.

Blanco-Gutiérrez, I., Varela-Ortega, C., & Flichman, G. (2011). Cost-effectiveness of groundwater conservation measures. A multi-level analysis. *Agricultural Water Management, 98*, 639–652.

BOE. (2006). Real Decreto 287/2006, de 10 de marzo, por el que se regulan las obras urgentes de mejora y consolidación de regadíos, con objeto de obtener un adecuado ahorro de agua que palie los daños producidos por la sequía. [Royal Decree 287/2006, of 10 March, laying down urgent works for improvement and consolidation of irrigation systems, in order to obtain adequate water savings that alleviates damage caused by drought]. Boletín Oficial del Estado Núm. 60, 11/03/2006 [In Spanish].

Camacho, E. (2012). La modernización de regadíos: Ahorro de agua versus incremento en el consumo de energía. Jornada técnica sobre coste energético y producción de energía en comunidades de regantes. [Modernization of irrigation systems: Water savings versus increase in energy consumption. Technical seminar on energy costs and energy production in irrigation districts]. FENACORE, Madrid, 28 March 2012 [In Spanish].

CEDEX. (2013). Proyecto I+D sobre Caracterización de medidas en materia de regadío que contribuyan a la consecución del buen estado de las aguas, e indicadores de eficacia y sostenibilidad. [R & D project on characterization of irrigation measures that contribute to the achievement of good water status. Indicators of efficiency and sustainability]. Centro de Estudios y Experimentación de Obras Públicas (CEDEX), Madrid [In Spanish].

CHG. (2010). Documento del taller territorial: Tramo Medio del Guadalquivir y Bajo Genil. Plan Hidrológico de la Demarcación del Guadalquivir [Document from the territorial workshop: Middle and Lower sections of rivers Guadalquivir and Genil. Hydrological Plan for the Guadalquivir Riverbasin]. Confederación Hidrográfica del Guadalquivir. Retrieved from http://www.chguadalquivir.es/export/sites/default/portalchg/planHidrologicoDemarcacion/demarcacionGuadalquivir/primerCicloPlanificacion/planHidrologicoGuadalquivir2009-2015/talleresJornadas/ficheros/taller11_documento2_documento_trabajo.pdf [In Spanish].

Corominas, J. (2010). Agua y energía en el riego, en la época de la sostenibilidad [Water and energy in irrigation, in the era of sustainability]. *Ingeniería del Agua, 17* (3), 219–233 [In Spanish].

Díaz Barcos, A. (2001). Beneficios derivados de la modernización de regadíos. El ejemplo de Caparroso (Navarra). *Navarra Agraria*, 46–50. [In Spanish].

Dumont, A., Mayor, B., & López-Gunn, E. (2013). Is the rebound effect or Jevons paradox a useful concept for better management of water resources? Insights from the irrigation modernization process in Spain. At the confluence - selection from the 2012 world water week in Stockholm. *Aquatic Procedia, 1*, 64–76.

Ederra, I. (2010). Eficiencia energética en el regadío [Energy efficiency in irrigation]. Jornada INTIC de interés profesional sobre innovación e ingeniería del agua: nuevas oportunidades y tendencias, 8 April 2010. Madrid [In Spanish].

ESTEPA (2010). Los regadíos históricos españoles. Paisajes culturales, paisajes sostenibles [Spanish historical irrigation. Cultural landscapes, sustainable landscapes]. Hermosilla, J. (Director). Unidad de Investigación Estudios del Territorio, del Paisaje y del Patrimonio. Universidad de Valencia y Ministerio de Medio Ambiente y Medio Rural y Marino [In Spanish].

Fernández García, I., Rodríguez Díaz, J. A., Montesinos, P., Camacho, E., & Berbel, J. (2013). La modernización de regadíos: repercusiones sobre el uso del agua y la energía en CCRR de Andalucía [The modernization of irrigation: Impact on water use and energy in irrigation communities of Andalusia]. XXXI Congreso Nacional de Riegos (AERYD), Orihuela (Alicante), 18–20 June 2013 [In Spanish].

Garrido, A., & Calatrava, J. (2009), Agricultural water pricing: EU and Mexico, OECD, Paris. Retrieved from http://www.oecd.org/dataoecd/25/38/45015101.pdf

Hanson, B. R. (1995). Practical potential irrigation efficiencies. First International Conference on Water Resources Engineering, San Antonio, TX, 14–18 de Agosto de 1995.

Hardy, L., Garrido, A., & Juana, L. (2012). Evaluation of Spain's water-energy nexus. *International Journal of Water Resources Development, 28*, 151–170.

Huffaker, R., & Whittlesey, N. (2003). A theoretical analysis of economic incentive policies encouraging agricultural water conservation. *International Journal of Water Resources Development, 19*, 37–53.

Hussain, I., Turral, H., Molden, D., & Ahmad, M. D. (2007). Measuring and enhancing the value of agricultural water in irrigated river basins. *Irrigation Science, 25*, 263–282.

Jensen, M. E. (2007). Beyond irrigation efficiency. *Irrigation Science, 25*, 233–245.

Junta de Andalucía. (2002). Inventario y caracterización de los regadíos de Andalucía (1996–2002). [Inventory and characterization of irrigation systems in Andalusia (1996–2002)]. Consejería de Agricultura y Pesca de la Junta de Andalucía. Seville [In Spanish].

Krinner, W., García, A., & Estrada, F. (1994). Method for estimating efficiency in Spanish irrigation systems. *Journal of Irrigation and Drainage Engineering, 120*, 979–986.

Lecina, S., Isidoro, D., Playán, E., & Aragües, R. (2010a). Irrigation modernization and water conservation in Spain: The case of Riegos del Alto Aragón. *Agricultural Water Management, 97*, 1663–1675.

Lecina, S., Isidoro, D., Playán, E., & Aragüés, R. (2010b). Irrigation modernization in Spain: Effects on water quantity and quality—A conceptual approach. *International Journal of Water Resources Development, 26*, 265–282.

López-Cortijo, I. (2008). Una visión general de los equipamientos en parcela, como obras de interés agrícola privado [An overview of the equipment within the plots of land, such as works of private agricultural interest]. Jornada técnica sobre pivot, Estación de Viticultura y Enología de Navarra (EVENA), Olite (Navarra), 29 April 2008 [In Spanish].

MAGRAMA. (2010). Análisis de la economía de los sistemas de producción. Resultados técnico-económicos de explotaciones agrícolas de Andalucía en 2009 [Analysis of the economics of production systems. Technical and economic results of farms in Andalusia in 2009]. Ministerio de Agricultura, Alimentación y Medio Ambiente. Madrid [In Spanish].

MAGRAMA. (2012). Plan Nacional de Regadíos [National Irrigation Plan]. Ministerio de Agricultura, Alimentación y Medio Ambiente, Madrid. Retrieved from http://www.magrama.gob.es/es/agua/temas/gestion-sostenible-de-regadios/plan-nacional-de-regadios/texto-completo/default.aspx [In Spanish]

MAGRAMA. (2015). Encuesta sobre Superficies y Rendimientos de Cultivos. Informe sobre Regadíos en España. 2014. Ministerio de Agricultura, Alimentación y Medio Ambiente. Madrid. Retrieved from http://publicacionesoficiales.boe.es/ [In Spanish].

MARM. (2010). Informe de viabilidad del Proyecto de modernización de la zona regable del Genil Margen Izquierda 2ª Fase, T.M. de Palma del Río (Córdoba). Clave CO-3116 [Feasibility report of the Project for the modernization of the irrigation area of Genil Left Bank 2nd Phase, in Palma del Río (Cordoba). Key CO-3116]. Ministerio de Medio Ambiente y Medio Rural y Marino, Madrid [In Spanish] Retrieved from http://www.magrama.gob.es/es/agua/planes-y-estrategias/informes-de-viabilidad-de-obras-hidraulicas/reggenilf_tcm7-27292.pdf.

Mateos, L., Lozano, D., Baghil, A. B. O., Diallo, O. A., Gómez-Macpherson, H., Comas, J., & Connor, D. (2010). Irrigation performance before and after rehabilitation of a representative, small irrigation scheme besides the Senegal river. *Agricultural Water Management, 97*, 901–909.

MIMAM. (2007). El agua en la economía española: situación y perspectivas. Informe integrado del análisis económico de los usos del agua en España. Artículo 5 y Anejo III de la Directiva Marco de Agua [Water in Spanish economy: situation and prospects. Integrated report of economic analysis of water uses in Spain. Article 5 and Annex III of the Water Framework Directive]. Ministerio de Medio Ambiente (2007), Madrid. Retrieved from https://circabc.europa.eu/webdav/CircaBC/env/wfd/Library/framework_directive/implementation_documents_1/wfd_reports/member_states/spain/article_5/Completo_Usos_nivel1.pdf

MMA. (2000). Plan Hidrológico Nacional. Análisis económicos [Spanish Hydrological Plan. Economic analyses]. Ministerio de Medio Ambiente, Madrid [In Spanish]. Retrieved from http://www.cedex.es/NR/rdonlyres/3BCDDD86-EC2B-441C-8C38-933EDD8B5E31/130031/PHN_MT_V5.pdf

Montero, J., & Tarjuelo, J. M. (2004). Analysis of water application cost with permanent set sprinkler irrigation systems. *Irrigation Science, 23*, 103–110.

OECD (2008). *Environmental performance of Agriculture in OECD countries since 1990*. París: OCDE. Retrieved from www.oecd.org/tad/env/indicators.

Perry, C., Steduto, P., Allen, R. G., & Burt, C. M. (2009). Increasing productivity in irrigated agriculture: Agronomic constraints and hydrological realities. *Agricultural Water Management, 96*, 1517–1524.

Peterson, J. M., & Ding, Y. (2005). Economic adjustments to groundwater depletion in the high plains: Do water-saving irrigation systems save water? *American Journal of Agricultural Economics, 87*, 147–159.

Pfeiffer, L., & Lin, C.-Y. C. (2012). Groundwater pumping and spatial externalities in agriculture. *Journal of Environmental Economics and Management, 64*, 16–30.

Playán, E., & Mateos, L. (2006). Modernization and optimization of irrigation systems to increase water productivity. *Agricultural Water Management, 80*, 100–116.

Playán, E., Slatni, A., Castillo, R., & Faci, J. M. (2000). A case study for irrigation modernization: II Scenario análisis. *Agricultural Water Management, 42*, 335–354.

Plusquellec, H. (2009). Modernization of large-scale irrigation systems: Is it an achievable objective or a lost cause? *Irrigation and Drainage, 58*, S104–S120.

Rodríguez Díaz, J. A., Camacho, E., & Blanco, M. (2011a). Evaluation of water and energy use in pressurized irrigation networks in Southern Spain. *Journal of Irrigation and Drainage Engineering, 137*, 644–650.

Rodríguez Díaz, J. A., Camacho, E., López, R., & Pérez, L. (2008). Benchmarking and multivariate data analysis techniques for improving the efficiency of irrigation districts: An application in Spain. *Agric Syst, 96*, 250–259.

Rodríguez Díaz, J. A., Pérez, L., Camacho, E., & Montesinos, P. (2011b). The paradox of irrigation scheme modernization: More efficient water use linked to higher irrigation demand. *Span J Agric Res, 9*, 1000–1008.

Rodríguez Ferrero, N., Salas, M., & Sánchez, M. T. (2010). Assessment of productive efficiency in irrigated areas of Andalusia. *Intl J Water Res Devel, 26*, 365–379.

Rogers, D. H., Lamm, F. R., Alam, M., Trooien, T. P., Clark, G. A., Barnes, P. L., & Mankin, K. (1997). *Efficiencies and water losses of irrigation systems*. MF-2243. Kansas State University, Research and Extension.

Scherer, T., & Weigel, J. (1993). Planning To Irrigate... A Checklist. NDSU Extension Service, 1993. North Dakota State University, EE UU.

Schierling, S. M., Young, R. A., & Cardon, G. E. (2006). Public subsidies for water-conserving irrigation investments: Hydrologic, agronomic, and economic assessment. *Water Resour Res, 42*, W03428.

Semaan, J., Flichman, G., Scardigno, A., & Steduto, P. (2007). Analysis of nitrate pollution control policies in the irrigated agriculture of Apulia region (Southern Italy): A bio-economic modelling approach. *Agric Syst, 94*, 357–367.

SWSI (2004). *Colorado water conservation board*. Colorado (USA): Statewide Water Supply Initiative. Department of Natural Resources.

Tirado, D., Gómez, C. M., & Lozano, J. (2006). Efficiency improvements and water policy in the Balearic Islands: A general equilibrium approach. *Investigaciones Económicas, XXX*, 441–463.

Torres, M. (2004). Avances técnicos en la Desalación de aguas [Technical advances in water desalination]. Revista Ambienta, Octubre 2004 (pp. 17–26). Ministerio de Medio Ambiente Rural y Marino, Madrid [In Spanish].

Turner, K., Georgiou, S., Clark, R., Brower, R., & Burke, J. (2004). Economic valuation of water resources in Agriculture. From the sectorial to a functional perspective of natural resource management. *FAO Water Reports No. 27*. Rome, Italy: FAO. Retrieved from ftp://ftp.fao.org/agl/aglw/docs/wr27e.pdf

Ward, F. A., & Pulido-Velázquez, M. (2008). Water conservation in irrigation can increase water use. *Proceedings of the National Academy of Sciences, 105*, 18215–18220.

Zanou, B., Kontogianni, A., & Skourtos, M. (2003). A classification approach of cost effective management measures for the improvement of watershed quality. *Ocean Coast Manage, 46*, 957–983.

Drivers of groundwater use and technical efficiency of groundwater, canal water, and conjunctive use in Pakistan's Indus Basin Irrigation System

Dawit Mekonnen[a], Afreen Siddiqi[b] and Claudia Ringler[a]

[a]Environment and Production Technology Division, International Food Policy Research Institute, Washington, DC, USA; [b]Massachusetts Institute of Technology, Cambridge, USA

ABSTRACT

This paper explores the major determinants of heavy reliance on groundwater and the extent to which conjunctive use of ground and surface water affects the production efficiency of Pakistan's irrigators. The results show that the major drivers of groundwater use in Pakistan's agriculture are the variability and uncertainty associated with surface water delivery and that any effort to address the groundwater–energy nexus challenge should first consider fixing the problems associated with surface water supplies. The findings also suggest that having access to groundwater does not directly translate into improvements in technical efficiency of production.

Introduction

Food security in Pakistan – a country that hosts the world's sixth-largest population, more than 180 million – is intimately tied with efficient national agricultural production. Historically, agricultural production in the Indus Basin of Pakistan has relied on a vast surface network of canals that spread water from the Indus and its tributaries over large tracts of land. This system – established during British rule in India – was designed to protect against famine rather than to maximize productivity (Jurriens, Mollinga, & Wester, 1996). With growing need for increased production, surface water supplies have been increasingly augmented with pumped groundwater. The irrigation intensity in some parts of this region is now among the highest in the world, with more than 75% of agricultural land under irrigation (Siebert, Portmann, & Döll, 2010).

In addition to expansion of irrigation, fertilizer use has increased substantially and new seed varieties have been introduced to enhance agricultural productivity, improve food security and reduce rural poverty (Hussain, Mudasser, Hanjra, & Amrasinghe, 2004). These measures have led to some gains in increasing crop yields, though productivity gaps persist across the country. Punjab is the major agricultural producer among the provinces in Pakistan, with a reported share of 76% for wheat, 73% for sugar cane, and 82% for cotton in total national production (PDS, 2013). At the district (sub-provincial) scale, however, productivity varies

Supplemental data for this article can be accessed at http://dx.doi.org/10.1080/07900627.2015.1133402.

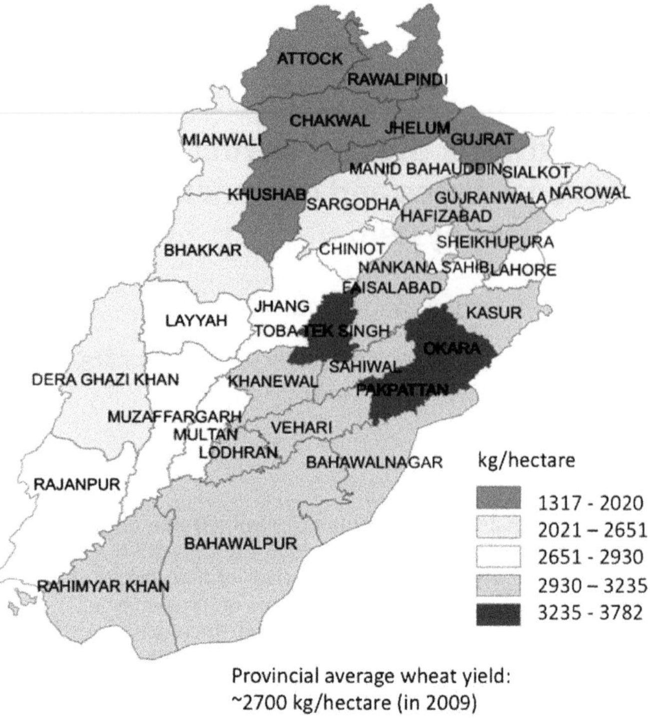

Figure 1. Wheat yield in districts of the Punjab province of Pakistan (data shown for 2008–2009). Provincial average wheat yield was approximately 2700 kg/ha in 2009. Data source: PDS (2012).

considerably (Figure 1). Yields of wheat (an important staple in the region) have increased threefold in some districts over the past three decades, whereas in other districts there has been stagnation or decline. Furthermore, the productivity gaps are also apparent in international comparisons. The maximum wheat yield in 2008–2009 was 3782 kg/ha in Pakistan's Punjab Province (PDS, 2012), compared to 4330 kg/ha in neighboring Indian Punjab (Chawla, Khepar, Sondhi, & Yadav, 2010).

Understanding the yield gap and factors affecting productivity has been an important question for socio-economic development of rural areas in Pakistan, and the country's land and water productivity has been studied in depth over the past several decades. Previous studies have estimated the productivity of irrigation water using regression analysis on farm survey data, and found that returns on irrigation water were high relative to costs of obtaining water (Hussain & Young, 1985). A micro-level analysis of groundwater markets in two districts in Pakistan indicated that all types of irrigation water – canal, purchased groundwater, and owned tubewell water – increased yields of wheat. However, groundwater had higher impact on yields than canal water, and water from tubewells owned by a farmer provided higher impact on yield (due to greater degree of control over irrigation) as compared to purchased groundwater (Meinzen-Dick, 1996). The variation in productivity across all the different canal commands has also been analyzed, and the results showed a wide range with a ratio of 1:5 in gross value of production per cubic metre of water in Punjab (Tahir & Habib, 2001). Within districts, there are further variations in productivity due to differences

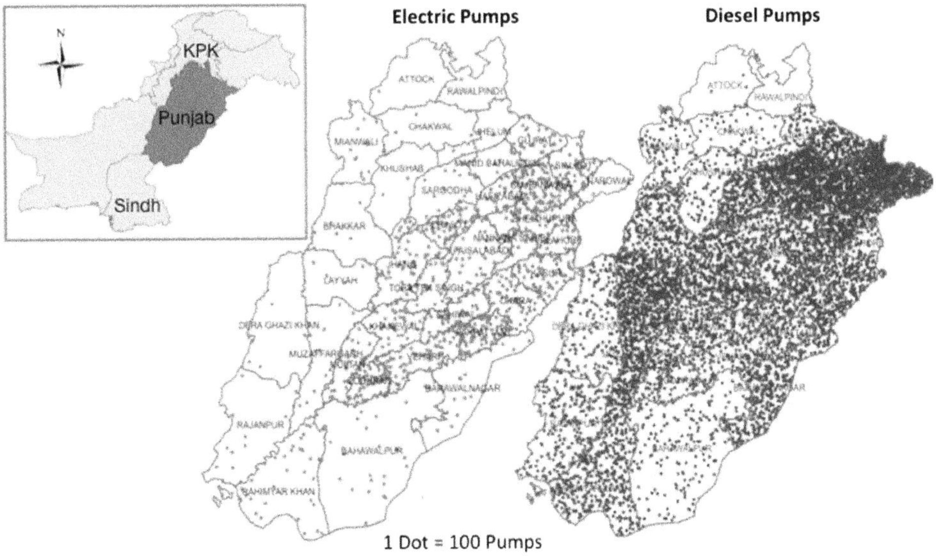

Figure 2. Electric and diesel pump installations in Punjab as of 2008–2009. Data source: PDS (2012).

in access to surface irrigation water. According to Qureshi, McCornick, Sarwar, and Sharma (2010), farms at the head of a canal receive 32% more water than farms at the tail end.

The previous work has provided important insights on the factors that have led to the emergence of a conjunctive irrigation system in Pakistan, but no work has been undertaken recently to investigate factors influencing current groundwater use. This article analyzes the current major drivers of reliance on groundwater use, followed by an investigation of the impact of conjunctive irrigation (use of both surface and groundwater) on production efficiency of wheat, using data from the Pakistan Rural Household Panel Survey conducted in 2011–2012 (International Food Policy Research Institute/Innovative Development Strategies [IFPRI/IDS], 2012). In the context of an ever-increasing reliance on groundwater use in Pakistan in the last two decades, with its consequences (increased energy demand for water extraction and application, and reduced soil health through increased salinity), this study identifies the major determinants of use of groundwater as entry points for policies aimed at addressing the groundwater–energy problem.

Literature review and background

The large-scale use of groundwater for agriculture is prevalent in many regions of the globe (Gleeson, Wada, Bierkens, & van Beek, 2012). In Pakistan, groundwater pumping started in the 1960s with a public-sector program to reduce the rising water tables in the Indus Basin due to seepage of freshwater from the canal system. In the 1970s, the government provided active support for expanding access to groundwater by paying capital installation costs and subsidizing electricity (Qureshi et al., 2010).

In the 1980s, with continued growth in pumping, the effects of over-abstraction started to become apparent. The government introduced regulations in the 1990s, including a licensing system to restrict installation of private tubewells in areas where groundwater tables were

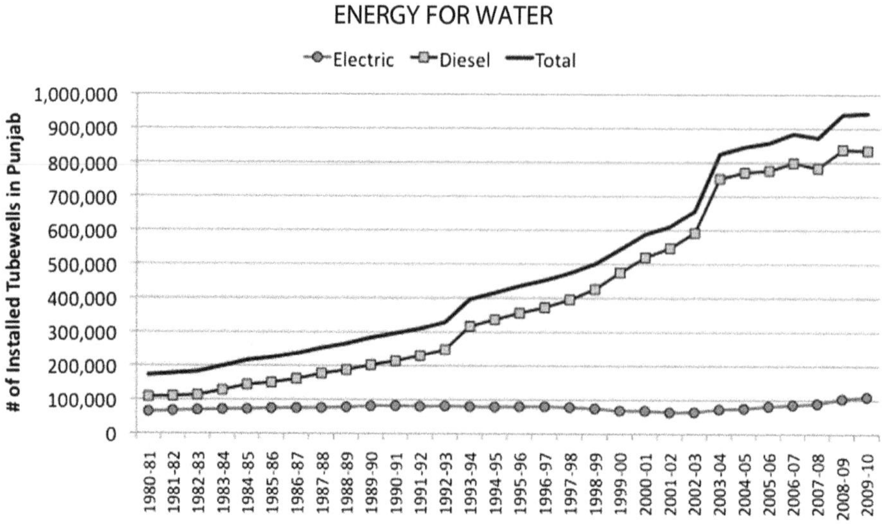

Figure 3. Total number of diesel and electric tubewells installed in Punjab. Source: MoWP (2012).

rapidly falling, and at the provincial level worked with the World Bank to prepare a groundwater regulatory framework for Punjab, the largest agricultural province in the country. In addition to direct management (through licensing), indirect approaches of energy pricing have also been applied, to little effect. The electricity tariff subsidy was removed in Punjab and Sindh in order to arrest the growth in groundwater pumping. The removal of electricity subsidies, together with a growing number of brownouts due to poor national energy policy, resulted in a shift to diesel pumps, which provided continuous access (no power cuts), were mobile and suitable for use in the increasingly fragmented land areas (due to shrinking farm sizes), and had lower installation costs. Electric tubewells were mostly used by large landowners, since the installation costs of obtaining a new electric connection from a main power line (including the costs of step-down transformers and cables) was prohibitive for small farmers (Qureshi, Shah, & Akhtar, 2003; Meinzen-Dick, 1996).

Currently, tubewells using diesel pumps constitute the major type of groundwater pumping in the country, accounting for 88% of total groundwater pumping in Punjab in 2011 (PDS, 2013). Diesel-based pumping puts appreciable demands on the energy sector in Pakistan – a country that is a net importer of petroleum products – with a reported average production of 64,000 barrels per day and consumption of 437,000 barrels per day in 2013 (EIA-Pak, 2015). The crude oil imports strain the national economy, and fuel shortages (including petrol and diesel) have grown (BS, 2015). Figure 2 shows the installation density of electric and diesel pumps in the various districts of Punjab Province.

A number of factors have been identified as contributing to growth in groundwater pumping. Hussain and Young (1985) report high returns from irrigation water (including from canal water and some groundwater pumping) relative to cost, explaining exploration of groundwater when canal water was insufficient. Various studies report that farmers pump groundwater because it allows for greater control and on-demand access, whereas canal water is unreliable and not available on demand (Qureshi et al., 2003; Kazmi, Ertsen, & Asi, 2012; Kumar, Scott, & Singh, 2013). The Indus Basin Irrigation System functions as a fixed surface supply–based system in which canal water is delivered at particular time intervals to farms. This time-based roster allocation system – known as *warabandi* – has been the mechanism of water distribution among farmers in the region (Briscoe & Qamar, 2006). Due to the unreliability of surface water supplies, however, farmers tend to over-irrigate when canal water becomes available, reducing crop yield and quality. Kumar and van Dam (2013) have also explained that in many developing countries, heavily subsidized and inefficient pricing of surface water, along with a close-to-zero marginal cost of electricity to pump groundwater, has created a situation where the private cost of marginal increase in the use of both water and electricity in agriculture is almost zero, reducing the incentive for farmers to invest in water-saving technologies.

Farmers have also resorted to increased groundwater use in times of drought. During 2000–2001, surface water availability was only 26% of 1996 levels (Qureshi et al., 2003). The drought continued for two years (Xie, Ringler, Zhua, & Waqas, 2013), followed by an almost 50% increase in new groundwater pump installations (Figure 3).

Another important factor has been population growth and associated land fragmentation, which has led to some farmland being cut off from direct access to canal water. Particularly smaller farmers often do not have access to canal water supplies, and use groundwater for irrigation (Qureshi et al., 2003).

In Punjab, groundwater is used on approximately 77% of irrigated areas, either alone or in conjunctive use with canal water (PDS, 2013). The pumping systems have provided critical access to irrigation water but have also increased the energy intensity of agricultural production in the region (Siddiqi & Wescoat, 2013). During 2001–2012, the price of light diesel oil increased from Rs. 16/L to Rs. 91/L (PEY, 2012). The electricity tariff for the agricultural sector also increased. For instance, the Multan Electric Power Company (serving districts with some of the highest agricultural electricity use in the province) had a fixed monthly rate of Rs. 70/KW in 2004, which increased to Rs. 120/KW in 2012. The variable rate increased from Rs. 2.75/kWh to 6.77/kWh during this period (PEPCO, 2013). Despite the rising energy costs (and therefore increased costs for pumping, which makes groundwater a very expensive source compared to almost-free canal water), groundwater abstractions continued to increase, indicating persistent challenges in the canal water delivery system.

This study investigates the factors driving growth in tubewell installations, as well as the impact of relying on canal water, groundwater and conjunctive use of canal and ground water as sources for irrigation, on technical efficiency of wheat production. The results inform irrigation management in this and other large conjunctive-use systems. Furthermore, they help identify entry points for energy policy reform in irrigated agriculture and policies to improve production efficiencies.

Empirical model

The empirical investigation first explores the major drivers of reliance on groundwater use, followed by an investigation of the implications of excessive reliance on groundwater vis-à-vis conjunctive use of ground and surface water on production efficiency. The next two sections explain the specific models used in the empirical approach to answer these two sets of questions.

Drivers of groundwater use

Farmers use groundwater either as an exclusive source of irrigation or as a supplement to canal water because of water shortages in the canal system at the time farmers need to irrigate their fields. But such reasoning does not directly lend itself to specific policy recommendations on how to solve shortages of both canal and groundwater. This article formally shows that water shortages are the main reason forcing farmers to use groundwater, but more importantly it identifies characteristics of canal water supply that could lead to mismanagement, aggravating shortages, and hence to increasing demand for groundwater use.

Factors that determine groundwater use can be (1) farm-specific, such as the location of the farm on a watercourse, salinity levels, waterlogging, slope, and fertility of the plots; (2) watercourse-specific, such as the location of the watercourse on the distributary, whether the watercourse is lined, whether the watercourse has a water users' association (WUA), number of farmers on the watercourse, number of tenants on the watercourse and its ratio to land owners, participation of some landlords of the watercourse in the irrigation department or other government offices, whether farmers exchange canal turns on the watercourse, land-holding distribution in the watercourse, etc.; (3) distributary-specific, such as the number of weeks the distributary or minor is open during the season, number of weeks the canal is full or half-full in the weeks the canal is officially open for the season, and whether the canal has a farmer organization (FO) that reports to the irrigation department; and (4) farmer-specific, such as household demographics and size of land holdings.

We model the reliance on groundwater (g_i) as an outcome of interest with a value of 1 if the farmer identified groundwater as the primary source of irrigation on the farm plot, and 0 otherwise. We also use tubewell ownership as an alternative outcome variable. We use the following logistic model, which includes factors that determine farmers' decisions to use groundwater as the primary source of irrigation water (or to own tubewells):

$$P(g_i = 1) = \frac{\exp(\alpha_0 + \alpha_H H + \alpha_F F + \alpha_W W + \alpha_D D)}{1 + \exp(\alpha_0 + \alpha_H H + \alpha_F F + \alpha_W W + \alpha_D D)} \quad (1)$$

where $P(g_i = 1)$ is the probability that groundwater is the main source of irrigation on the farm (or the farmer owns a tubewell); *H, F, W* and *D* are vectors of household, farm, watercourse and distributary characteristics as described above; and α_0, α_H , α_F, α_W, and α_D are corresponding parameters to be estimated.

Implications of groundwater, surface water, and conjunctive use on technical efficiency of production

A stochastic frontier production function is estimated for wheat production in Punjab, Sindh and Khyber Pakhtunkhwa (KPK) Provinces of Pakistan to explore the extent to which conjunctive use of surface and groundwater, or lack thereof, affects the technical efficiency of farms. We estimate the technical efficiency of irrigated agriculture in Pakistan and quantify the contribution of varying irrigation practices as well as household and plot characteristics to average production inefficiency.

Wheat was chosen for the analysis because of its dominance in the *rabi* (winter) season, accounting for more than 70% of cultivated land during this season (IFPRI/IDS, 2012). In addition, the focus on wheat (and the *rabi* season) is helpful to abstract from crop choice issues and the resulting crop-specific water requirements in the more diversified *kharif* (summer) season.

Variables that are expected to represent the production environment are modeled such that they will influence the mean technical inefficiency of farmers following the stochastic frontier model developed by Aigner et al. (1977) and extended by Battese and Coelli (1995) and Wang (2002). The basic formulation is:

$$y_i = x_i\beta + v_i - u_i \qquad (2)$$

for households $i = 1, \ldots, N$, where y_i is the natural log of wheat production per unit of land for household i; x_i is a vector of the log of production inputs (fertilizer, capital, pesticide, labor, and volume of irrigation water – both surface and groundwater); v_i is a zero-mean random error, assumed to be independently and identically distributed as $N(0, \sigma_v^2)$; and u_i is a non-negative random variable associated with the technical inefficiency of production, which measures the percentage of output lost due to inefficiency and is assumed to be distributed as truncated normal $N^+(u, \sigma_u^2)$. The technical inefficiency component of the error term, u_i, is expressed as $u_i = f(z_i\delta)$, where z_i is a vector of variables thought to explain inefficiency, such as the relative reliance of the household on groundwater compared to surface water, and other controls that can affect technical efficiency. These control variables include soil fertility levels, average slope of the plot, soil types, WUAs, farms' locations on the watercourse, whether the watercourse is lined, location of the watercourses on the distributary that feeds the watercourse, number of weeks the distributary is officially open for the season, number of weeks the distributary was half or less than half full, presence of a distributary-level FO, tenancy status of the plot, problems of waterlogging and salinity, and household head's age and education; δ is a vector of associated coefficients to be estimated.

The technical efficiency score of farm i is computed as $TE = \exp(-u_i)$. The production function component, y_i, and the inefficiency effects, u_i, are estimated together in one step. The model is estimated for wheat production during the *rabi* season of 2011–2012 using the Pakistan Rural Household Panel Survey (Round 1.5), which provides detailed plot-level information with highly disaggregated data on irrigation types, methods and institutions (IFPRI/IDS, 2012).

The production function is specified as a Cobb-Douglas production function. Due to the extensive list of variables included in the model, translog specification with interaction terms of the variables was not tractable to converge. District-specific dummy variables are included in the production function component, while province-level dummy variables are included among the factors that explain inefficiency instead of the district dummies due to high collinearity of the district dummies with distributary- and watercourse-level variables. Of the 942 farm households in the survey, 561 farms were found that produced wheat in the 2011–12 *rabi* season. However, linking farmers' plots with watercourse and distributary names on which we have detailed information from a community survey was possible only for about 467 farms. We chose to use the smaller sample size to get more insights on watercourse- and distributary-level information. For this reason, the sample is not nationally representative. Nevertheless, it provides necessary insights on drivers of groundwater use and the impact of conjunctive use on production efficiency.

Results

Reliance on groundwater use is captured in two ways: whether groundwater is the primary source of irrigation water for the plot, and whether there is a tubewell on the farm land.

Table 1. Determinants of using groundwater as a primary source for irrigation in wheat production in Punjab, Sindh, and Khyber Pakhtunkhwa Provinces of Pakistan.

	(1)	(2)
Explanatory variables	OLS coefficients	Logit average marginal effects
Location on the watercourse: middle	0.261***	0.319***
	(0.060)	(0.080)
Location on the watercourse: tail	0.135**	0.209**
	(0.067)	(0.082)
Tenancy: own	0.023	0.023
	(0.045)	(0.041)
Use precision land leveling	−0.043	−0.042
	(0.072)	(0.080)
Total size of land ownership	−0.001	−0.001
	(0.004)	(0.004)
Household head attended school	−0.065*	−0.059*
	(0.037)	(0.034)
Family size	−0.002	−0.002
	(0.005)	(0.004)
Age of the household head	−0.001	−0.001
	(0.001)	(0.001)
Watercourse lined	0.001*	0.001*
	(0.001)	(0.001)
Water users' association in the watercourse	−0.196***	−0.232***
	(0.058)	(0.059)
Someone in the watercourse works for the irrigation department or is a politician	−0.185***	−0.116**
	(0.058)	(0.059)
Farmers sell canal turns on the watercourse	0.215***	0.163**
	(0.062)	(0.064)
Watercourse located at the tail of the distributary (compared to middle)	0.172***	0.221***
	(0.066)	(0.069)
Number of weeks the distributary is officially open in the season	−0.026***	−0.036***
	(0.004)	(0.007)
Number of weeks the canal was half or less than half full in the season	0.003	0.008**
	(0.004)	(0.004)
Distributary-level farmer organization	−0.284***	−0.182***
	(0.055)	(0.049)
Sindh	−0.563***	−0.470***
	(0.090)	(0.053)
KPK	−0.110*	−0.206***
	(0.063)	(0.051)
Constant	0.755***	
	(0.127)	
Observations	467	467
R^2	0.494	

Standard errors in parentheses.
$^{***}p < .01$; $^{**}p < .05$; $^{*}p < .1$.

We present results from these alternative definitions of reliance on groundwater in wheat production in Punjab, Sindh, and KPK Provinces of Pakistan in Tables 1 and 2. In both cases, results from ordinary least squares (OLS) estimation and logit estimations are presented for comparison. However, the interpretation of findings focuses on the results from the preferred logit specification given the binary values of the dependent variables. The results from the logit specification refer to average marginal effects of the explanatory variables on changes in the predicted probabilities of the dependent variable. That is, at least for categorical variables, the marginal effects can readily be interpreted as the impact of changes in an explanatory variable on the probability of using groundwater as a primary source for

Table 2. Determinants of tubewell ownership in wheat production in Punjab, Sindh, and Khyber Pakhtunkhwa Provinces of Pakistan.

	(1)	(2)
Explanatory variables	OLS coefficients	Logit average marginal effects
Location on the watercourse: middle	0.129*	0.188**
	(0.068)	(0.074)
Location on the watercourse: tail	0.008	0.076
	(0.077)	(0.077)
Tenancy: own	−0.015	−0.028
	(0.052)	(0.047)
Use precision land leveling	−0.125	−0.130*
	(0.083)	(0.077)
Total size of land ownership	0.025***	0.026***
	(0.004)	(0.004)
Household head attended school	−0.014	−0.007
	(0.042)	(0.041)
Family size	0.003	0.002
	(0.005)	(0.005)
Age of the household head	−0.001	−0.000
	(0.002)	(0.002)
Watercourse lined	0.000	0.000
	(0.001)	(0.001)
Water users' association in the watercourse	−0.110*	−0.094
	(0.066)	(0.066)
Someone in the watercourse works for the irrigation department or is a politician	−0.188***	−0.177***
	(0.066)	(0.060)
Farmers sell canal turns on the watercourse	0.011	0.008
	(0.071)	(0.074)
Watercourse located at the tail of the distributary (compared to middle)	−0.023	−0.016
	(0.076)	(0.077)
Number of weeks the distributary is officially openin the season	−0.008*	−0.007*
	(0.004)	(0.004)
Number of weeks the canal was half or less than half full in the season	−0.013***	−0.012***
	(0.005)	(0.005)
Distributary-level farmer organization	0.010	0.017
	(0.063)	(0.064)
Sindh	−0.516***	−0.385***
	(0.103)	(0.036)
KPK	−0.223***	−0.180***
	(0.072)	(0.059)
Constant	0.511***	
	(0.145)	
Observations	467	467
R^2	0.250	

Standard errors in parentheses.
$^{***}p < .01$; $^{**}p < .05$; $^{*}p < .1$.

irrigation or on the probability of tubewell ownership. Using the definition of groundwater as the primary source of water fits the data well, with an R^2 of 0.49 (Table 1), compared to using the definition of tubewell ownership, with an R^2 of 0.25 (Table 2).

The results in Column 2 of Table 1 indicate that the major driver of groundwater use in Pakistan is the shortage and unreliability of surface water delivery. Farms at the middle and tail of a watercourse are 32% and 21% more likely, respectively, to have groundwater as a primary source of irrigation compared to farms at the head of a watercourse. The higher probability for farms at the middle of a watercourse suggests a degree of complementarity between surface and groundwater use, possibly to dilute saline groundwater with sweet surface water. In addition to the location of the farms on a watercourse, the location of the

Table 3. Elasticities of wheat yield to changes in production inputs in the stochastic frontier production function.

Dependent variable (y): log of wheat yield (kg/acre)	
Explanatory variables (x)	Elasticity
Log of fertilizer used (kg/acre)	0.033**(0.015)
Log of machinery hours used per acre	0.007(0.007)
Log of number of sprays used per acre	0.001(0.001)
Log of labor days used (days/acre)	0.016(0.017)
Log of total water used (inches)	0.013**(0.006)
District fixed effects controlled for?	Yes
Observations	333

Standard errors in parentheses.
*$p < .10$; **$p < .05$; ***$p < .01$.

Table 4. Combination of groups used in the analysis based on source of irrigation water and location on watercourse.

	Groundwater	**Canal water + groundwater**	**Canal water**
Head	Group 1	Group 2	Group 5
Middle	Group 1	Group 4	Group 7
Tail	Group 1	Group 3	Group 6

watercourse on the distributary or minor that feeds the watercourse is an important determinant of reliance on groundwater use. Farms on watercourses at the tail of a distributary are 22% more likely to use groundwater as a primary source of irrigation compared to farms on watercourses at the middle of a distributary.[1] Similarly, an additional week the minor or distributary is officially open in the season reduces the probability of using groundwater as a primary source of irrigation by 3.6%, while an additional week the distributary is half or less than half full increases the probability of using groundwater as a primary source of irrigation by about 1%. All these results indicate that any effort to address the groundwater–energy nexus challenge should first consider the main cause of increased pumping, which is the insufficient and unreliable canal water supply.

The results in Table 1 also show that groundwater is less likely to be a primary source of irrigation on watercourses and distributaries with formally organized WUAs. Farms on watercourses with WUAs are 23% less likely to use groundwater as a primary source of water, while farms on distributaries with distributary-level FOs that report to the irrigation department are 18% less likely to use groundwater as a primary source of water for irrigation. These results suggest that WUAs and FOs could be strengthened and used as effective institutions to improve canal service delivery, and hence reduce excessive reliance on groundwater.

The results in Table 1 also provide evidence for significant differences across provinces in the use of groundwater in Pakistan. Farms in Sindh are 47% less likely to use groundwater as a primary source of irrigation, compared to farms in Punjab. Likewise, farms in KPK are 21% less likely to use groundwater as a primary source of water, compared to farms in Punjab. The difference across provinces in their reliance on groundwater as a primary source of water could be the result of differences in both groundwater potential and groundwater quality across the provinces in favor of Punjab. The total available groundwater resources of Punjab, Sindh, and KPK Provinces are estimated at 42.75 million, 18 million, and 3.11 million acre-feet, respectively (PPSGWP, 1998).[2] Similarly, groundwater quality is better in Punjab,

Table 5. Marginal effects on the average of technical inefficiency.

	Marginal effect on mean technical inefficiency, *u*	
Explanatory variables	Average marginal effect	Standard deviation
Source of irrigation water and location on the watercourse		
Household at the head of the watercourse using only canal water	−0.160	0.101
Conjunctive user of ground and canal water located at the head of the watercourse	−0.047	0.030
Household at the middle of the watercourse using only canal water	−0.190	0.119
Conjunctive user of ground and canal water at the middle of the watercourse	−0.103	0.065
Conjunctive user of ground and canal water at the tail of the watercourse	0.035	0.022
Household at the tail of the watercourse using only canal water	−0.250	0.157
Watercourse characteristics		
Water users' association on the watercourse	−0.068	0.043
Watercourse lined	0.002	0.001
Watercourse located at the tail of the distributary (compared to middle)	0.078	0.049
Distributary characteristics		
No. of weeks the distributary is officially open in the season	−0.010	0.006
No. of weeks the canal was half or less than half full in the season	0.003	0.002
Distributary-level farmer organization	0.154	0.097
Plot characteristics		
Flat land	−0.062	0.039
Slightly sloped land	−0.010	0.006
Very fertile	−0.067	0.042
Sandy soil	0.149	0.094
Sandy loam soil	0.159	0.100
Loam soil	0.026	0.017
Tenancy: own	0.015	0.009
Plots experience salinity	−0.021	0.013
Plots experience waterlogging	−0.097	0.061
Farmer characteristics		
Household head attended school	0.023	0.015
Log of age of household head	−0.005	0.003
Location variables		
Sindh	0.620	0.390
KPK	0.234	0.358
Observations	333	

Note: The dependent variable is technical inefficiency, so positive signs correspond to exacerbated inefficiency, and negative signs to improved technical efficiency.

compared to Sindh: a relatively smaller share of the canal command areas in Punjab (23%) have hazardous groundwater quality, while the share is 78% in Sindh (Qureshi et al., 2010).

The presence of landowners on the watercourse who work for the irrigation department or who are politicians reduces the probability of using groundwater as a primary source of water by about 11%. We argue that political connections of farm owners increase the chance of getting canal water in watercourses on their farms, reducing the relative reliance on groundwater.

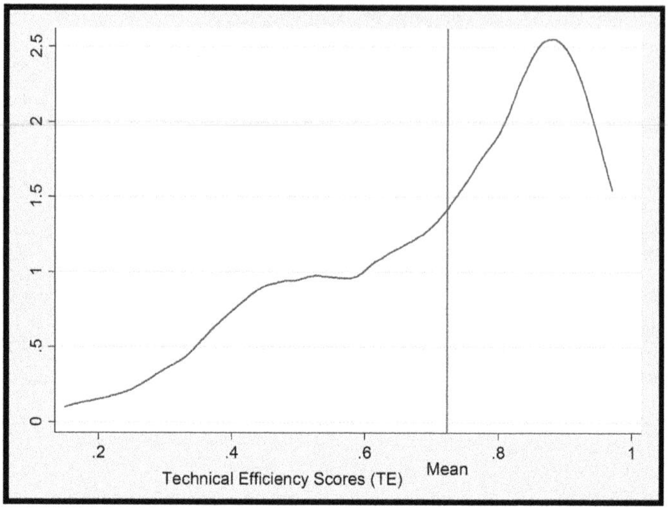

Figure 4. Kernel density of technical efficiency scores of wheat farmers in the sample. Source: authors' computation.

Watercourse lining has a positive relationship with groundwater use. Increasing canal lining of a watercourse by 10% increases the probability of using groundwater as a primary source of water by about 1%. This could be because of the difficulty of tampering with lined watercourses, forcing anyone who needs additional water to get it from groundwater. On the other hand, lined watercourses could also indicate irrigation office strategies to keep canal water separate from saline groundwater. In some places in Pakistan's irrigation system, saline groundwater is applied jointly with sweet canal water to mitigate absolute surface water shortages.

Farms with educated household heads are 6% less likely to use groundwater as a primary source of water. The survey data suggest that educated household heads have somehow managed to get more canal water from the system: they reported about 20% more frequent irrigations using canal water, reducing the need to rely on groundwater as a primary source of water (IFPRI/IDS, 2012). This does not imply that educated households are less likely to use groundwater. On the contrary, the survey data suggest that educated households use more of both canal and groundwater. However, they are less likely to use groundwater as a *primary* source of water. It is possible that better education, which is generally associated with a higher knowledge base, seems to facilitate farmers' ability to navigate the canal water delivery system to position themselves with relatively better access to canal water.

Farms on watercourses that sell canal turns to each other are 16% more likely to use groundwater as a primary source of water. Descriptive statistics show that canal water shortage seems severe in watercourses where farms sell canal turns to each other. For instance, 40% of the farmers on watercourses that do not sell canal turns report that the timing of the canal turn is as per their needs; this falls to only 11% in watercourses that sell canal turns (IFPRI/IDS, 2012). The data also show that the volume of canal water applied to wheat production in watercourses that do not sell canal turns is three times the volume of canal water in watercourses that sell canal turns. Thus, the informal canal water markets are indicators of the severe water scarcity in these specific canals, forcing farms to rely more on groundwater.

Results from using an alternative definition of reliance on groundwater in terms of tubewell ownership are presented in Table 2. Farms in the middle of a watercourse are 19% more likely to own a tubewell than farms at the head of a watercourse. The positive effect of tail-end users on tubewell ownership compared to farms at the head is not significantly different from zero. Ownership of an additional acre of land[3] increases the likelihood of tubewell ownership by 2.6%, possibly because the surface water delivery may be insufficient for large holdings, or because large farms can support the cost of tubewell installations more easily, or both. In addition, having landlords on the watercourse who work for the irrigation department or are politicians reduces the probability of tubewell ownership by 18%.

Irrigation canals and distributaries in Pakistan can be perennial (ensuring regular supply of water year-round), non-perennial (running in monsoon and summer), or inundation (running only during rainy season). The volume of these canals can also vary within a season. Thus, the number of weeks the distributary is open for the season as well as the level of the canal in the weeks it is open can influence farms' reliance on groundwater. The results show that an additional week the distributary is officially open in the season reduces the probability of tubewell ownership by about 0.7%. Likewise, an additional week the distributary is at least half full in the season reduces the probability of tubewell ownership by 1.2%. These results suggest that improving canal water delivery should be an integral part of groundwater management in Pakistan.

Farms that use precision land leveling are 13% less likely to own tubewells compared to farms that do not use land leveling. Farms in Sindh and KPK Provinces are 39% and 18% less likely, respectively, to own tubewells compared to farms in Punjab Province of Pakistan (Table 2).

The full set of estimated coefficients from the stochastic production frontier is presented in Table A1 in the online supplementary data (available at http://dx.doi.org/10.1080/07900627.2015.1133402). For ease of interpretation and analysis, the elasticities of wheat yield with respect to changes in the production inputs are presented in Table 3, while the marginal effects of the explanatory variables on average technical inefficiency scores are presented in Table 5, even though the production function and factors determining technical inefficiency are estimated together in one step as suggested in the efficiency literature. Wheat yield is found to be highly responsive to additional volume of irrigation water and to increases in fertilizer applications (Table 3). A 1% increase in the amount of irrigation water applied leads to a 1.3% increase in wheat yield. This elastic response of wheat yield to water suggests that yield from the last incremental unit of irrigation water is higher than the average yield obtained from all units of irrigation water[4]. A 1% increase in the amount of fertilizer used per acre increases wheat yield by about 3.3% (Table 3).

Technical efficiency is measured for a given level of input use; it measures how much more a farmer can produce at the same level of input use or, conversely, how much input can be reduced without reducing the current level of production. Conjunctive use of groundwater and canal water, or lack thereof, is one of the main variables affecting the technical inefficiency of farmers. It is hypothesized that this effect depends on the location of the farms on the watercourse, and hence farmers are classified into different groups based on water source and location on the watercourse. Table 4 depicts the different combinations and the groups used in the model.

As shown in Table 4, the different farmer groups are: (1) farmers who use only groundwater; (2) conjunctive users of ground and canal water at the head of a watercourse; (3)

conjunctive users of ground and canal water at the tail of a watercourse; (4) conjunctive users of ground and canal water at the middle of a watercourse; (5) farmers at the head of a watercourse using only canal water; (6) farmers at the tail of a watercourse using only canal water; and (7) farmers at the middle of a watercourse using only canal water.

The marginal effect of determinants of technical inefficiency on the average inefficiency score, computed using the method suggested by Wang (2002), are presented in Table 5. Any one of the seven groups in Table 4 could be used as comparison group; Table 5 shows the results when farms using no canal water were used as a comparison group.

Farmers who use only canal water for irrigation obtain much higher technical efficiency than those who use only groundwater, in head, middle and tail reaches of the canal (16%, 19%, and 25%, respectively). Further, the farmers who practise conjunctive use in middle reaches of the canal are 10% more efficient than those who use only groundwater (Table 5).

On the other hand, conjunctive use of ground and surface water does not necessarily lead to improvements in the efficiency of wheat production among farms at similar relative positions on the watercourse. There is no statistically significant difference in the efficiency of farms at the head of a watercourse with and without groundwater use. Similarly, there is no difference in the efficiency of wheat production for farms at the middle of a watercourse with and without groundwater. Conjunctive water users at the tail of the watercourse are even less technically efficient (by about 23 percentage points) than farms using only canal water at the tail. This is possibly due to the inferior quality of groundwater impacting production negatively in reducing yield or due to under-application of water as a result of groundwater shortages or as a result of high energy cost. A study focused on the Chaj sub-basin in Pakistani Punjab found that groundwater quality varied significantly across reaches in canal command areas and had significant adverse impacts on yields. Wheat production was highly profitable with only canal water use, and least profitable with only groundwater use (Hussain et al., 2004). Similarly, Kumar, Trivedi, and Singh (2009), who compared yield for major crops in the Bist Doab area of Indian Punjab across different sources of irrigation water, reported that wheat yield is higher (by about 10%) for canal water compared to groundwater. For at least one of the study locations, the authors argued that the better yield response of wheat to canal water could be due to better quality of the canal water in the area.

These results are also consistent with the findings of Kumar, Singh, and Sivamohan (2010) that farmers who pay higher prices for irrigation water (such as groundwater users) use it more efficiently by allocating more land and water away from wheat and rice and towards crops that give higher returns per unit of land. For instance, Kumar and van Dam (2013) showed that pomegranate grown in the north Gujarat region of India gives a net return of USD 1.8/m^3 of water, against USD 0.08/m^3 for wheat, while Kumar et. al. (2010) reported that net return for diesel pump owners in the Uttar Pradesh region of India is two to three times lower for wheat, and four to seven times lower for paddy, compared to crops such as sugar-cane, potato, pearl millet, pea, gram, mustard and linseed. As a result, and particularly when faced with growing water scarcity in semi-arid and arid regions, well irrigators are allocating more water to highly water-efficient crops (Kumar & van Dam, 2013; Kumar et al., 2010; Kumar, 2005). Thus, the lack of technical inefficiency in wheat production in our study is more likely a reflection of groundwater users' response to higher energy costs and water scarcity by directing more of their water and other inputs (including managerial oversight) away from wheat and towards other crops that provide more economic returns per unit of water or land.

It is also important to note that the virtues of groundwater over canal water in terms of control, reliability, and on-demand access, are likely to be offset by higher energy prices, due to the heavy reliance of groundwater on energy inputs, while almost all of the canal water delivery in Pakistan uses gravity irrigation. As a result, when energy prices rise, groundwater users are more likely to curtail the frequency and duration of irrigation compared to those who use canal water, with the resulting negative impact on yield and technical efficiency measures.

Given the amount of water they use, canal-only water users at the middle and tail of watercourses are 19% and 25% more efficient, respectively, than farms that use ground and surface water at the head. Similarly, canal-only water users at the tail of a watercourse are 33% more efficient than conjunctive water users in the middle of a watercourse, while conjunctive water users at the tail are 27% less efficient than conjunctive water users in the middle of a watercourse. The general trend suggests that farmers who face worse surface water shortages are more likely to use their resources as efficiently as possible, while farms with relatively easier access to water (such as those at the head or those that supplement surface water with groundwater) are found to be less technically efficient in terms of producing more for a given level of input.

The results also show that farms on sandy and sandy-loam soils are less efficient than the comparison group of farms with clay soils in wheat production. This is consistent with water leaving the root zone more quickly through sandy soils, so insufficient moisture is present. An additional week in which the distributary or minor is officially open for the season reduces technical inefficiency (and hence increases production) by 1%. Farms in Sindh and KPK Provinces are 62% and 23% less efficient, respectively, than farms in Punjab. However, farms on distributaries or minors with distributary-level FOs are less efficient than farms on distributaries without such institutions. In addition, farms on lined watercourses are associated with lower technical efficiencies, though the difference is small.

The results of the stochastic frontier model show that the average technical efficiency score for wheat production in Pakistan is 72% (Figure 4). Thus, there is a potential to increase wheat yield by about 39% (28/72) through improved management of existing levels of input use. This average technical efficiency score is comparable with efficiency scores reported by other researchers for wheat production in Pakistan. Battese, Malik, and Gill (1996) reported efficiency scores in four districts of Punjab ranging from 57% to 79%, while Ahmad, Chaudhry, and Iqbal (2002) reported a technical efficiency score of 68% among sampled farmers in Punjab, Sindh, and KPK.

Findings and conclusions

Pakistan witnessed a gradual increase in the use of groundwater resources starting in the 1960s and 1970s, when the government actively supported large-scale use of groundwater to reduce the rising water tables in the Indus Basin due to seepage from the canal system. However, since the mid-1990s, farmer-owned tubewells and pumps have proliferated at an exponential rate in the country, and this growth has led to problems of overdraft, falling water tables, and degradation of groundwater quality (Qureshi et al., 2010). In the context of continued large groundwater abstraction, this study used household survey data to analyze the major determinants of groundwater use, and the implication of groundwater use for farms' technical efficiency of wheat production.

The findings suggest that the major drivers of groundwater use in Pakistan's agriculture are the variability and uncertainty associated with surface water delivery. Groundwater is more likely to be the primary source of irrigation water for farms at the middle or tail of watercourses, and farms on watercourses at the tail of distributaries, on distributaries that are officially open for relatively fewer weeks, and on those distributaries for which the volume of water is half or less than half full for a considerable number of weeks in a season. All these results indicate that any effort to address the groundwater–energy problem should first consider the main cause of increased pumping, which is the insufficient and unreliable canal water supply.

The results also show that groundwater is less likely to be a primary source of irrigation on watercourses and distributaries with formally organized WUAs and FOs at watercourse and distributary or minor levels. These results suggest that WUAs and FOs could be strengthened and used as effective institutions to improve canal service delivery and reduce excessive reliance on groundwater.

The findings also suggest that complete reliance on groundwater significantly reduces the technical efficiency of wheat production in Pakistan. Moreover, conjunctive use of ground and surface water does not necessarily lead to improvements in the efficiency of wheat production. For a given location of farms along a watercourse, either there is no statistically significant difference in the technical efficiency of farms with and without conjunctive groundwater use, or when there is a difference, such as at the tail of watercourses, it is in favor of farms that rely only on canal water. These results are likely due to (1) inferior quality of groundwater impacting productivity negatively, (2) under-application of water as a result of groundwater shortages or as a result of high energy costs, and (3) the relatively higher cost of inefficiency on farms that face higher surface water shortages (such as those at the tail of watercourses), putting pressure on them to use their resources as efficiently as possible compared to farms with relatively easier access to water.

Overall, the results show that any effort to address the problems of groundwater abstraction and associated energy consumption should focus on the main driver of increased pumping, which is unreliable canal water supply. Fixing the problems of the canal system, be it by reducing conveyance losses in the primary, secondary and tertiary canals; by improving the governance of the canal system, such as reflected in the substantial efficiency improvements brought about by the watercourse-level WUAs; or by increasing reservoir storage capacity to reduce the uncertainties in the supply of surface water, could play an important role in reducing groundwater pumping. Improved canal water management, despite its role in reducing reliance on groundwater pumping, may not translate into improved groundwater balances, because watercourse lining or improved application efficiency will reduce return flows from the canal system. The reality of increasing energy costs of groundwater pumping, however, requires that better canal management be an integral component of the policy mix in addressing the groundwater–energy challenge.

In summary, this empirical analysis has identified important factors that influence the groundwater economy of Pakistan, and sets the stage for further work. In future work, more research is needed to investigate why access to groundwater does not readily translate into productivity and efficiency improvements. It would be important to determine whether quality differences between surface and groundwater sources are impacting productivity, and whether related environmental stresses, such as salinity problems from excessive groundwater use, are key factors suppressing yield gains.

Notes

1. Only 2% of the households in the sample were on watercourses at the head of a distributary; these are combined with the middle location on the distributary.
2. One acre-foot is 1233.48 m^3.
3. An acre of land is equivalent to 0.405 ha.
4. This is because elasticity can be shown to be a ratio of marginal product of an extra input and average product of the total input used so far.

Acknowledgements

The authors acknowledge the insights and comments received from the editor and two anonymous referees. Any and all errors are the sole responsibility of the authors.

Disclosure statement

No potential conflict of interest was reported by the authors.

Funding

This research was supported with generous funding from the CGIAR Research Program on Water, Land, and Ecosystems and from the Pakistan Strategy Support Program under the CGIAR Research Program on Policies, Institutions, and Markets.

References

Ahmad, M., Chaudhry, G. M., & Iqbal, M. (2002). "Wheat productivity, efficiency, and sustainability: A stochastic production frontier analysis". *The Pakistan Development Review, 41* Part II (Winter 2002), 643–663.

Aigner, D., Knox Lovell, C. A., & Schmidt, P. (1977). Formulation and estimation of stochastic frontier production models. *Journal of Econometrics, 6*, 21–37.

Battese, G. E., & Coelli, T. J. (1995). A model of technical inefficiency effects in a stochastic production function for panel data. *Empirical Economics, 20*, 325–332.

Battese, G. E., Malik, S. J., & Gill, M. A. (1996). An investigation of technical inefficiencies of production of wheat farmers in four districts of Pakistan. *Journal of Agricultural Economics, 47*, 37–49.

Briscoe, J., & Qamar, U. (Eds.) (2006). *Pakistan's water economy: Running dry*. Washington, DC: The World Bank.

BS. (2015), Business standard news, "Pakistan shaken by fuel crises". URL: Retrieved from March, 2105. http://www.business-standard.com/article/news-ians/pakistan-shaken-by-fuel-crisis-115011800353_1.html

Chawla, J. K., Khepar, S. D., Sondhi, S. K., & Yadav, A. K. (2010). Assessment of long-term groundwater behaviour in Punjab, India. *Water International, 35*, 63–77. doi:10.1080/02508060903513502

EIA-Pak. (2015). *US Energy Information Administration*, URL; Retrieved from Mar, 2015. http://www.eia.gov/countries/country-data.cfm?fips=pk.

Gleeson, T., Wada, W., Bierkens, M. F. P., & van Beek, L. P. H. (2012). Water balance of global aquifers revealed by groundwater footprint. *Nature, 488*, 197–200.

Hussain, I., Mudasser, M., Hanjra, M. A., & Amrasinghe, U. (2004). Improving wheat productivity in Pakistan: Econometric analysis using panel data from chaj in the upper indus basin. *Water International, 29*, 189–200.

Hussain, R. Z., & Young, R. A. (1985). Estimates of the economic value productivity of irrigation water in Pakistan from farm surveys. *Water Resources Bulletin, 21*, 1021–1027.

International Food Policy Research Institute/Innovative Development Strategies. (2012). *Pakistan Rural Household Panel Survey 2012, Round 1.5, [Dataset]*. Washington, D.C./Islamabad, Pakistan: IFPRI/IDS. http://dx.doi.org/10.7910/DVN/T9GGYA

Jurriens, M., Mollinga, P. P., & Wester, P. (1996). *Scarcity by design: Protective irrigation in India and Pakistan, paper 1*. Wageningen Agricultural University, International Institute for Land Reclamation and Improvement.

Kazmi, S. I., Ertsen, M. W., & Asi, M. R. (2012). The impact of conjunctive use of canal and tubewell water in Lagar irrigated area, Pakistan. *Physics and Chemistry of the Earth, 47–48*, 86–98.

Kumar, M. D. (2005). Impact of electricity prices and volumetric water allocation on energy and groundwater demand management: analysis from Western India. *Energy Policy, 33*, 39–51.

Kumar, M. D., Scott, C. A., & Singh, O. P. (2013). Can India raise agricultural productivity while reducing groundwater and energy use? *International Journal of Water Resources Development, 29*, 557–573.

Kumar, M. D., Singh, O. P., & Sivamohan, M. V. K. (2010). Have diesel price hikes actually led to farmer distress in India? *Water International, 35*, 270–284. doi:10.1080/02508060.2010.484910

Kumar, M. D., Trivedi K., & Singh O. P. 2009. Analyzing the impact of quality and reliability of irrigation water on crop water productivity using an irrigation quality index. In M. Kumar, and U. Amarasinghe (Eds.), *Strategic analyses of the National River Linking Project (NRLP) of India, series 4. Water productivity improvements in Indian agriculture: potentials, constraints and prospects* (pp. 55–72). Colombo, Sri Lanka: International Water Management Institute.

Kumar, M. D., & van Dam, J. C. (2013). Drivers of change in agricultural water productivity and its improvement at basin scale in developing economies. *Water International, 38*, 312–325.

Meinzen-Dick, R. (1996). *Groundwater markets in Pakistan: participation and productivity*. Washington, D.C.: International Food Policy Research Institute. Research Report 105

MoWP. (2012), *Handbook on water statistics of Pakistan*. Islamabad: Ministry of Water and Power.

PDS. (2012). *Punjab development statistics - 2012*. Punjab: Bureau of Statistics.

PDS. (2013). *Punjab development statistics - 2013*. Punjab: Bureau of Statistics.

PEPCO. (2013). Pakistan electric power company DISCO (Distribution Company) performance statistics, 2013. Islamabad: Author.

PEY. (2012). *Pakistan energy year book – hydrocarbon development institute of Pakistan*. Islamabad: Ministry of Petroleum and Natural Resources.

PPSGWP. (1998). *Annual Progress Report*. Lahore: Punjab Public Sector Ground Water Project. 1998.

Qureshi, A. S., McCornick, P. G., Sarwar, A., & Sharma, B. R. (2010). Challenges and prospects of sustainable groundwater management in the indus basin, Pakistan. *Water Resource Management, 24*, 1551–1569.

Qureshi, A. S., Shah, T., Akhtar, M. (2003). *The groundwater economy of Pakistan* – Working Paper 64. IWMI.

Siddiqi, A., & Wescoat, J. L. (2013). Energy use in large-scale irrigated agriculture in the Punjab province of Pakistan. *Water International, 38*, 571–586.

Siebert, S., Portmann, F. T., & Döll, P. (2010). Global patterns of cropland use intensity. *Remote Sensing, 2*, 1625–1643. doi:10.3390/rs2071625

Tahir, Z., and Habib, Z. (2001). "Land and water productivity: Trends across Punjab canal commands". Working Paper 14. International Water Management Institute.

Wang, H. (2002). Heteroscedasticity and non-monotonic efficiency effects of a stochastic frontier model. *Journal of Productivity Analysis, 18*, 241–253.

Xie, H., Ringler, C., Zhua, T., & Waqas, A. (2013). Droughts in Pakistan: A spatiotemporal variability analysis using the standardized precipitation index. *Water International, 38*, 620–631.

Energy for water utilization in China and policy implications for integrated planning

Xi Li[a], Jie Liu[a], Chunmiao Zheng[b,c], Guoyi Han[d] and Holger Hoff[d]

[a]Institute of Water Sciences and College of Engineering, Peking University, Beijing, China; [b]School of Environmental Science and Engineering, South University of Science and Technology of China, Shenzhen, China; [c]Department of Geological Sciences, University of Alabama, Tuscaloosa, Alabama, USA; [d]Stockholm Environment Institute, Sweden

ABSTRACT

Significant energy resources are needed for traditional and nontraditional water utilization in China. Yet the interlinkages between water and energy have not received adequate attention in the country. To address this gap, this article disaggregates and quantifies the magnitude and direction of energy and water flows in China at the national level through Sankey diagrams. Spatial distributions of energy use by different components of the water supply were further mapped at the provincial level to discern regional differences. The results of this study show that the total energy consumption by water abstraction, treatment and distribution, as well as waste treatment and reuse, amounts to 193.5 TWh of electricity, or about 4% of the total national electricity usage. The outcome of this study offers important policy implications for integrated water and energy planning and management and will contribute to achieving the goal of low-energy water utilization in the future.

Introduction

China is a country with acute water problems caused by increasing demand, uneven spatial and temporal distribution of water resources, and water quality deterioration. These issues are exacerbating the country's water scarcity. Meanwhile, demand for both traditional and nontraditional water resources continues to rise, driven by the country's rapid economic expansion and industrialization. To meet growing water demand, energy used for extraction, transportation and treatment is increasing significantly. While projects such as deep groundwater pumping, long-distance water transfer and seawater desalination are adding crucial supplies of fresh water, they are consuming an ever-greater amount of energy. Although a good understanding of the water and energy interlinkages is important for integrated water and energy policy analysis and planning, study of the water–energy nexus in China is still in its infancy (Kahrl & Roland-Holst, 2008; Qin, Curmi, Kopec, Allwood, & Richards, 2015).

Supplemental data for this article can be accessed at http://dx.doi.org/10.1080/07900627.2015.1133403.

In this study, data from multiple sources are compiled and used to calculate energy consumption for different processes of water utilization in China, including water abstraction, treatment and distribution, as well as wastewater treatment and reuse. It is noteworthy that long-distance inter-basin water transfer projects that alter the water supply structure and the so-called Three Red Lines water management strategy, which mandates strict limits on water utilization and pollutant discharge, are two fundamental policy instruments being used by the Chinese government to cope with geographic imbalance in water resources and water shortage. A better understanding of water–energy linkages will provide valuable insights on these policy instruments and offer possible options for policy reforms involving water conservation strategies, water supply structure changes and technical solutions, which, in turn, will contribute to achieving the goal of low-energy water utilization in the future.

This article is organized as follows. It starts with an overview of energy and water resources in China through Sankey diagrams. It is followed by a methods and data section where the source of the data and method for calculation of energy use in various parts of the water sector are detailed. It is concluded with policy implications for integrated water-energy planning and management. Additional data and materials, including the distribution of average groundwater depths in China and an overview of China's major water transfer projects, are provided in the online supplemental information (available at http://dx.doi.org/10.1080/07900627.2015.1133403).

Overview of China's water and energy resources

Unless otherwise specified, the data used in this study are all from 2011. The choice is due to both data availability and the fact that 2011 is the starting year of the current Twelfth Five-Year Plan for National Socio-Economic Development in China. As a result, it offers a good reference year for a comparison with the end of the plan in 2015. Sankey diagrams are used to display water and energy flows in China from sources to destinations (Sankey, 1898; Schmidt, 2008), in which the arrow width is proportional to the related quantity.

Water utilization in China

According to the China Water Resources Bulletin (CWRB, 2012), China used about 610.7 km^3 of water in 2011 (excluding direct seawater use). Table 1 shows the share and flow of the water use, which was used to construct the Sankey diagram for water use in the country for that year (Figure 1). The water use is divided broadly into four categories: domestic; commercial and industrial; agricultural; and cooling (mainly for thermal power and nuclear power). In Table 1 and Figure 1, desalinated water is assumed to be used only by the commercial and industrial sectors, and harvested rainwater is assumed to be evenly distributed to the domestic and agricultural sectors. The surface water allocated to each water use category is calculated by subtracting other water supplies from the total water use. Water consumption rates from the China Water Resources Bulletin are used to calculate water consumed by different water users.

A few initial observations can be made about water use in China. First, a majority of the water supply comes from surface water and groundwater. Harvested rainwater (0.18%), recycled water (0.54%) and desalinated seawater (0.02%) combined supply only 1% of total freshwater used. The rate of recycled water use in China is far lower than in developed countries, which indicates a major potential. Second, groundwater in China provides about 40%

Table 1. The sources of data used for constructing Figure 1.

Categories	Source	Percentage used	Data sources
Water supply	Freshwater	Surface water (81%) Groundwater (18%) Others (1%)	CWRB*
	Seawater	direct use for cooling	
End users	Domestic (10%) Commercial/industry (22%) Agriculture (61%) Cooling (7%)		Calculated based on data from CWRB
Water use by sources	Groundwater	Domestic (20%) Commercial/industry (26%) (including 60% for cooling) Agriculture (54%)	Margat & Van der Gun (2013)
	Desalinated water	used by commercial and industrial only	Based on the current situation (SOA, 2013)
	Rainwater harvest	Domestic (50%) Agriculture (50%)	Based on the fact that the total amount of rainwater harvest is quite small (less than 1%) and the averaged distribution will have no significant impact on the results Wang et al. (2009); Li, Gong, & Wei (2000)
	Recycled water	Cooling (21%) Commercial/Industry (49%) Agriculture (30%)	Guo, Wang, and Zhou (2012)
	Surface water	Domestic (8%) Commercial/industry (24%) Agriculture (63%) Cooling (5%)	Calculated by subtracting other supply from the total water use. Other refers to other sources than surface water, including groundwater, desalinated and recycled water and rainwater harvest
Water return	Cooling	Discharge (93.5%) Consumed/evaporated (6.5%)	CWRB; NBS (2012c)
	Other water end users excluding cooling	Discharge (48%) Consumed/evaporated (52%)	Calculated using water consumption rate from CWRB CWRB

*CWRB: China Water Resource Bulletin.

of water for irrigation and 70% of drinking water in the northern and north-western regions (Qiu, 2010). Margat & Van der Gun (2013) have reported that 54% of groundwater is used for agriculture, 26% for industry and 20% for domestic use in China. China has become the second-largest groundwater user, behind India (Margat & Van der Gun, 2013). As the water table drops due to long-term over-exploitation of groundwater resources, one can expect each unit of water withdrawn to require more energy, in addition to the growing environmental and ecosystem impacts. Third, agriculture is the largest water user in China, accounting for

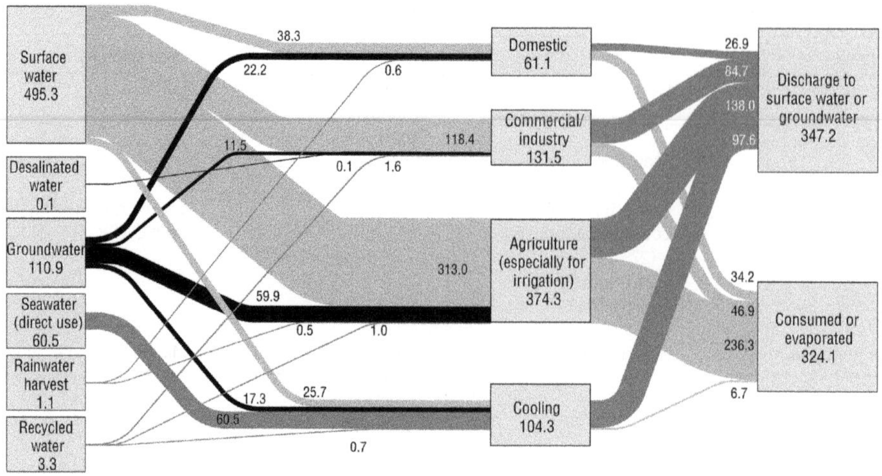

Figure 1. Sankey diagram of water flows from sources to destinations in China in 2011 (all amounts in km^3; sources listed in Table 1).

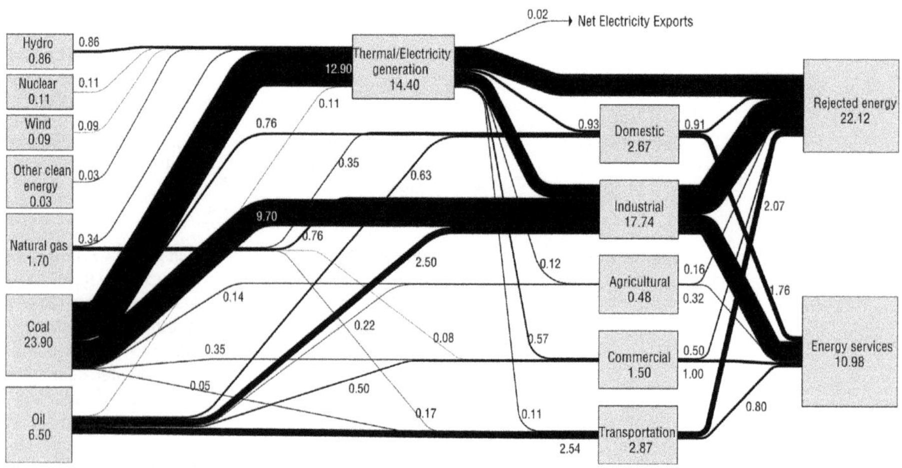

Figure 2. Energy use of China in 2011. All amounts are in tonnes of standard coal equivalent. The total energy used was approximately 3.31 × 10^9 TCE. Source: NBS (2012a).

roughly 61% of the total water withdrawals. Domestic, commercial and industrial water uses, which accounted for 31% of total water use in 2011, are rapidly growing. They are met by self-supply and public supply, in the proportion of 20% and 80%, respectively (MOHURD, 2012; NDRC, 2012). Lastly, cooling for thermal power annually consumes 7% of freshwater withdrawals and 60.5 km^3 of non-desalinated seawater directly.

Energy utilization in China

In 2013, China became the largest energy consumer in the world (U.S. Energy Information Administration, 2014). Previous studies have created a complex Sankey diagram of energy

Table 2. Energy intensity for various water processes in China and the data sources.

Item	Energy intensity (kWh/m^3)	Data sources
Surface water extraction	Water transfer, depending on different length of water transfer projects Water storage: 0.13 Gravity flow: 0 Water lifting from rivers and lakes: 0.20	Liu et al. (2013); Shao, Wang, & Wang (2003); Plappally & Lienhard V (2012); Gao (2012); Wen et al. (2014)
Groundwater extraction	Depending on average pump lift in each province (average is 0.37)	China Geological Environmental Monitoring Institute (2012); Wang et al. (2012)
Public water supply	0.30	China Urban Water Association (2012)
Desalinated water treatment	5.9	Al-Karaghouli & Kazmerski (2013); Buonomenna (2013); Gude (2011); Gude et al. (2010)
Recycled water treatment	0.82	Wen et al. (2014)
Waste water treatment	Provincial data (average is 0.26)	Fu & Zhong (2014)

flows from energy source to final consumption for the year 2005 (Ma, Allwood, Cullen, & Li, 2012). For a clearer and systematic understanding of energy utilization in China, this article constructs a relatively simple Sankey diagram for 2011, the year in which the corresponding water data was collected (Figure 2). The data were mainly derived from the 2011 energy balance sheet in the 2012 China Energy Statistical Yearbook (NBS, 2012a), which contains the energy data for supply, transmission and consumption by various users.

Figure 2 shows that the primary energy source in China is coal, which accounts for around 70% of the total energy supply. Roughly 50% of the coal is used to generate electricity, and the rest is used by industry. Oil is the second-largest energy source in China, accounting for 20% of the total energy supply. Interestingly, China was a net oil exporter before the early 1990s (U.S. Energy Information Administration, 2014) but since 2009 has become the world's second-largest importer of crude oil and petroleum products due to rapid industrialization. Clean energy, such as hydropower and wind, accounts for only 3% of the total energy supply based on calorific value calculation (NBS, 2012a). Rejected energy in Figure 2 consists primarily of heat losses, conversion losses and transmission losses as defined by the Lawrence Livermore National Laboratory (https://www.llnl.gov/). The overall energy use efficiency in China is only about 33%, or 10 percentage points lower than the average in developed countries (Wei & Shen, 2009).

Calculation of energy for water utilization: data and methods

Energy used by the water sector can be divided into constructional use and operational use (Rothausen & Conway, 2011). Constructional use refers to the energy used in constructing water-related infrastructure such as wells, conveyance pipes and treatment plants; operational use indicates the energy used to extract, convey, treat and distribute water. The former category is usually counted as energy used by industry and therefore is not considered in this study.

To obtain the energy intensity for all water-related processes described below, data were collected from Chinese government-authorized sources, as well as a series of reports and academic journal articles. The energy-intensity data used in this study, as well as their sources, are shown in Table 2 and explained in detail in the following sections.

Groundwater lifting

The high energy requirements for groundwater extraction are due to pumping, and vary a great deal depending on depth, type of pump, pumping efficiency, and volume of extraction. Concerning the groundwater depth, estimates are derived from the theoretical physical relationship between energy use and groundwater lifting (Equations (1) and (2)), which have been widely used in a number of studies (Li, Wang, & Qiu, 2013; Wang et al., 2012; Plappally & Lienhard V, Plappally & Lienhard, 2012). Equation (1) describes the energy required to lift one cubic metre of water a height of one metre assuming 100% efficiency – a figure calculated to be 0.00273 kWh (Rothausen & Conway, 2011). Equation (2) was used to calculate the total energy consumption based on Equation (1) and the actual groundwater withdrawals, efficiency, and pumping lift.

$$energy(kWh) = \frac{9.8\ N/kg \times pumping\ lift(m) \times mass(kg)}{3.6 \times 10^6 \times efficiency(\%)} \tag{1}$$

where *pumping lift* is the vertical distance from the water table to the ground (m) and *efficiency* is the overall pumping efficiency.

$$E_{Total} = \frac{0.00273 \times pumping\ lift(m) \times V\left(m^3\right)}{PE \times (1 - TDL)} \tag{2}$$

where *V* is volume of groundwater withdrawn (m^3), PE is pumping efficiency (including diesel and electric pumps), and TDL is transmission and distribution losses of the power supply system (only for electric pumps).

Equation (2) shows that pumping lift, pumping efficiency and pumped volume are the most important factors influencing energy requirements for groundwater extraction. Pumping lift is usually greater than groundwater depth because of the instantaneous drawdown in the well when pumping. It is difficult to obtain the accurate pumping lift in all pumping wells as well as the corresponding water withdrawn. Wang et al. (2012) proposed a correlation between the average groundwater depth and the pumping lift, which is adopted in this study. Groundwater-level data were compiled from observations published by the China Geological Environmental Monitoring Institute (2012) and field monitoring data collected by the authors. Average groundwater depths by provinces were spatially mapped (Figure A1 in the online supplemental data), and total groundwater withdrawals of each province were collected and analyzed. Figure A1 clearly indicates that intensive groundwater use is driving the decline of groundwater level. This 'aggregate' method of spatially averaging the groundwater depth was found to have no significant variation compared to methods that use individual pumping wells with actual pumping capacity (Bartos & Chester, 2014).

Pumping efficiency is another important factor for calculating energy consumption but has a large uncertainty due to the data availability at the national level. Diesel, electric and manual pumps are three major ways to pump groundwater in China, of which only the energy required for diesel and electric pumps is considered in this article. For the calculations in this study, the following steps were taken. First, groundwater withdrawn using manual pumps in each province was subtracted from total groundwater withdrawals. Data on manual groundwater extraction were obtained from the Bulletin of the First National

Table 3. A summary of inter-basin water transfer projects in China in 2011. Sources: Liu and Yang (2012); Liu et al. (2013).

Source	Destination	Quantity(km^3/y)	Average length of transfer(km)	Major water receiver
Yellow River	Hai River basin	3.95	342 (average of Yin Huang Ji Wei and Yin Huang Ji Jin projects)	Hebei Beijing, Tianjin
Yellow River	Huai River basin	2.72	200	Henan, Shandong
Yellow River	Shandong Peninsula	1.57	290	Qingdao Yantai Weihai
Yellow River	Inland river basin	0.26	–	Gansu
Yangtze River	Huai River basin	8.96	400	Jiangsu
Huai River	Yangtze River basin	0.49	400	Jiangsu

Census for Water (China National Census for Water, 2013). Second, for the remaining water, withdrawn using electric and diesel pumps, it was assumed that 76% were diesel while 24% were electric (Wang et al., 2012; Zou et al., 2013). Third, pumping efficiencies of diesel and electric pumps were estimated to be 15% and 40%, respectively, based on previous studies (Karimi, Qureshi, Bahramloo, & Molden, 2012; Shah, 2009; Wang et al., 2012). Lastly, transmission and distribution losses in the electric power network were estimated to be 7% (China Power, 2012).

Surface water conveyance pumping

The major energy requirement for surface water is conveyance. The energy required for surface water conveyance varies greatly with the type of hydraulic projects. According to the Ministry of Water Resources, there are basically four types of hydraulic projects: artesian water diversion; water lifting; water storage; and inter-basin water transfer, representing 33.7%, 32.3%, 30.4% and 3.6% of the total surface water withdrawals in China, respectively (CWRB, 2012). Artesian water diversion projects are designed to be gravity-driven and therefore do not require significant energy. Water lifting projects lift surface water from rivers or lakes. The energy required to lift a cubic metre of surface water one metre varies according to specific local conditions. Considering pump efficiency and transport distances based on survey data in previous studies in China, this article estimates the required energy to be 0.2 kWh/m^3 (Gao, 2012; Hu, Ou, Zhang, & Karplus, 2013). Water storage projects primarily refers to reservoirs. There is no specific estimate of energy intensity for reservoir water extraction in China. The transport distance of water from reservoirs is usually less than that from rivers and lakes, because the latter are always unevenly distributed in China. A previous study in the UK has also indicated that the energy required for surface water extraction from reservoirs is slightly lower than that from rivers despite construction costs (Reffold, Leighton, Choudhury, & Rayner, 2008). Considering this and survey data (Gao, 2012), this article assumes that the energy intensity for reservoir extraction is 0.13 kWh/m^3 in China, slightly lower than the number used for river and lake extraction. Inter-basin water transfer projects are the most energy-intensive type of surface water supply. The energy consumption of these projects depends primarily on the length of each individual transfer project. The energy use per unit distance for surface pipe flow is a function of the material or frictional resistance of the pipe, as well as the geometry of the pipeline (Plappally & Lienhard V, 2012). A previous study

Table 4. Seawater treatment plans by different technologies in China. Source: SOA (2013).

	Membrane technologies		Thermal technologies	
Plant type	Reverse osmosis	Electro- dialysis	Multi-stage flash distillation	Multiple-effect distillation
Plants	84	1	1	9
Desalination capacity percentage	60.97%	0.03%	0.77%	38.23%

calculated the energy use for water transfer projects per unit distance in China to be 0.0045 kWh/m^3 per km (Gao, 2012). Table 3 presents detailed information on the amount of water transfer among different river basins and the average lengths of the transfer projects (Liu & Yang, 2012; Liu et al., 2013).

Distribution and treatment

Depending on uses and quality requirements, freshwater can be either directly used or treated in public water plants before use. In the case of direct use, energy is required only to distribute the water from sources to users. In cases when treatment is necessary, more energy is required. Due to limited information at the national level, this study relies on an estimate of average energy intensity in public water plants of 0.3 kWh/m^3 (China Urban Water Association, 2012).

Seawater requires a much higher energy input for treatment before use. Table 4 shows that there were 95 seawater desalination plants in China in 11 coastal provinces at the end of 2012 (SOA, 2013). Some 75% of the plants are considered small or medium-sized (≤10,000 m^3/day), while 25% are large (>10,000 m^3/day). Table 4 shows that reverse osmosis and low-temperature multiple-effect distillation are the two major desalination technologies in China, and 61% of China's desalination capacity comes from reverse osmosis technology. In general, reverse osmosis is more energy-efficient than thermal processes. Reverse osmosis directly consumes electricity, while thermal processes are mainly driven by low-temperature steam extracted from an adjacent power plant with equivalent electrical energy consumption (Plappally & Lienhard V, 2012). In 2011, it is estimated that the average energy required to desalinate water in China was 5.9 kWh/m^3 (Al-Karaghouli & Kazmerski, 2013; Buonomenna, 2013; Gude, 2011; Gude, Nirmalakhandan, & Deng, 2010).

End users

The end users of water may be domestic, commercial/industrial, or agricultural. To determine different use-phase energy requirements, the water allocation from suppliers to each type of water user was examined. Detailed water flows are displayed in Figure 1, as discussed before. Agriculture is the major water user and consumes about 61% of the total water supply, which mainly comes from surface water and groundwater pumping, without requiring energy for water treatment. Apart from agriculture, the energy required for other types of water use is related not only to pumping but also to treatment.

ENERGY FOR WATER

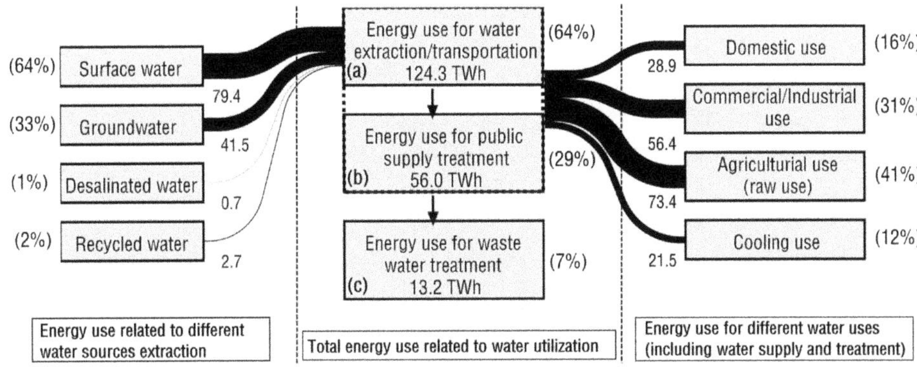

Figure 3. Energy use in the water sector in China in 2011.

Wastewater treatment

Wastewater treatment can be divided into two categories: preliminary treatment before discharging water to the environment, and high-standard treatment for water recycling. The amount of wastewater discharge in each province in 2011 was obtained from the China Statistical Yearbook (NBS, 2012c). Total wastewater discharge was 65.9 km^3, of which industrial wastewater and domestic sewage accounted for 35% and 65%, respectively. Guangdong, Jiangsu and Shandong are the three provinces with the largest wastewater discharge in China. The wastewater treatment rate in China in 2011 was around 80%, as reported in the China Statistical Yearbook on Environment (NBS, 2012b). Total recycled water in 2011 was only 3.3 km^3, less than 5% of the total treated wastewater. The recycling rate of water in China remains far lower than that of developed countries.

Energy intensity for preliminary wastewater treatment and recycled water treatment is influenced by various factors, such as regional differences, the capacity of the wastewater treatment plants, treatment technologies, and standards for processed wastewater. An increase in plant size usually results in a decrease in unit energy consumption. According to a report by the World Resources Institute (http://wri.org.cn/) that examined energy intensity for rough wastewater treatment in seven economically coordinated regions, average energy intensity for preliminary wastewater treatment in China in 2012 was about 0.26 kWh/m^3, slightly lower than the estimates for 2006 and 2009 (Fu & Zhong, 2014). For recycled water, the average energy intensity is 0.82 kWh/m^3 (Wen, Zhong, Fu, & Spooner, 2014).

It is noteworthy that there exists some inconsistency regarding the information on wastewater discharge and treatment rates, especially in rural areas. The related data from the China Statistical Yearbook are mainly for urban areas. In spite of this limitation, the data used in this article can still provide an overall reliable depiction of energy use for water sectors in China as a whole.

Results

Energy use in China's water sector

The Sankey diagram shown in Figure 3 depicts the energy required for water provision together with wastewater treatment in China in 2011. The chain of water utilization (including

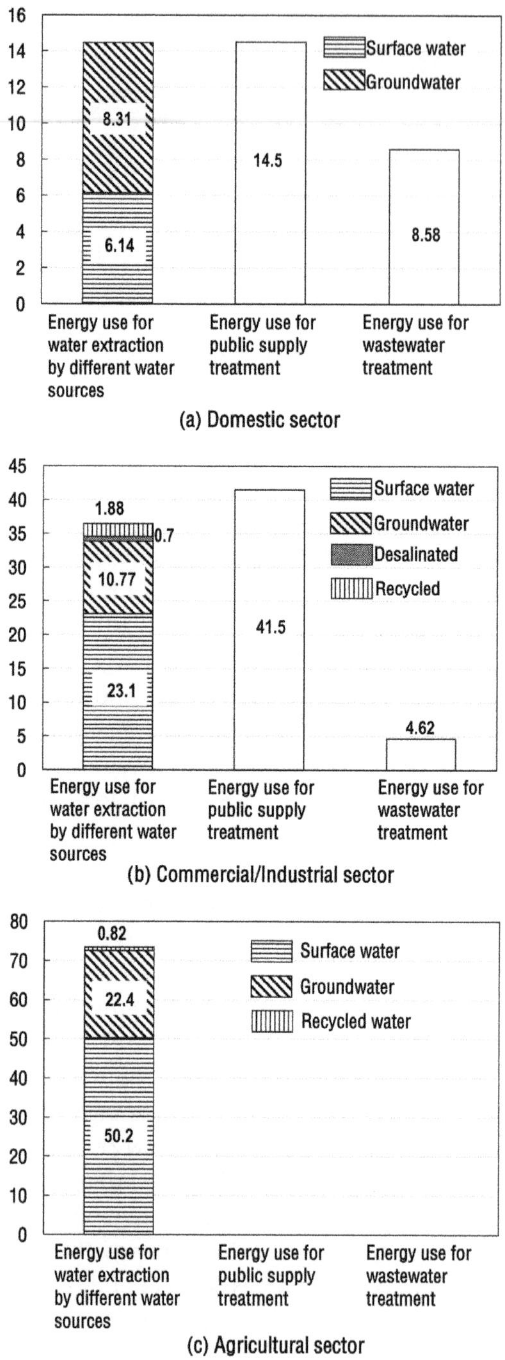

Figure 4. Energy use (in TWh) in each step of water provisioning and treatment for the domestic, agricultural and industrial sectors in China in 2011.

water extraction, treatment and distribution, as well as wastewater treatment and reuse) consumed 193.5 TWh (Figure 3), accounting for about 4% of total national electricity

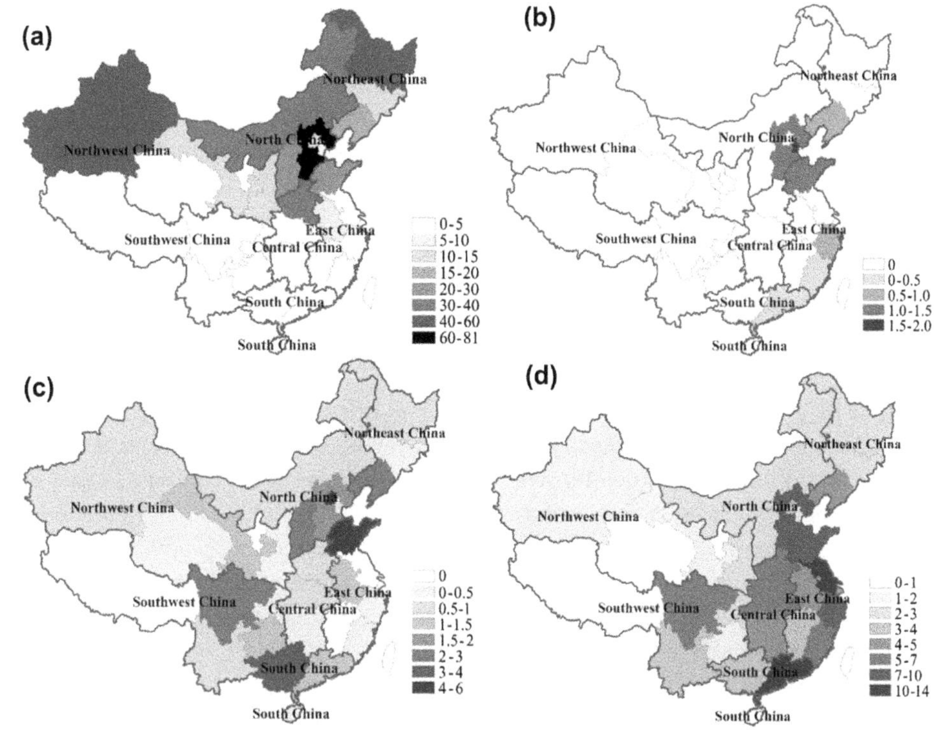

Figure 5. Provincial distributions of energy use for water in China in 2011 (in 10^8 kWh): (a) for groundwater extraction; (b) for desalination; (c) for water recycling; (d) for wastewater treatment.

consumption. The highest percentage of energy use was for water extraction (64%) and in particular the extraction of surface water (79.4 TWh). This is due to the large amount of surface water use and inter-basin water transfer projects. Energy use for wastewater treatment is relatively low due to the low wastewater treatment rate but has an increasing trend.

This work summed the energy used for water provision (including water extraction, transportation and treatment) for each water user, and the total is 180.3 TWh (Figure 3a, b). The energy for agricultural water provision contributes the highest proportion of the total energy use for water provision (41%), which is directly related to the large amount of water used for agriculture in China. Figure 4 shows the energy use by various water use sectors for different sources of water. Water utilization in the domestic sector consumed 37.5 TWh, including energy use for surface water and groundwater extraction (14.45 TWh), public water supply treatment (14.5 TWh), and wastewater treatment (8.58 TWh). As for the commercial/industrial sector (including cooling use), it consumed 82.5 TWh and was the largest energy consumer. The agricultural sector does not need energy for any water quality treatment; therefore it required only 73.4 TWh for water extraction and transportation.

Provincial distribution of energy use for water

The spatial distributions of provincial energy use for groundwater extraction, desalination, water recycling and wastewater treatment are shown in Figure 5. Various water-using structures result in different energy use for water utilization among provinces. Energy use for groundwater extraction is mainly concentrated in northern China due to the high proportion of groundwater use in those areas (Figure 5a). Desalinated water is used only in some coastal provinces at present, while recycled water, another nontraditional water resource, is more widely used around China (Figure 5b, c). Energy use for wastewater treatment has a linear relationship with the amount of wastewater discharge, which is higher in east China (Figure 5d).

Policy implications and challenges

Policy implications

Energy use by the water sector is directly proportional to the water use for prescribed technologies. Hence, water savings translate into energy savings, unless these water savings are achieved through more energy-intensive technologies. However, with China's rapid economic development, savings from water conservation will hardly be able to compensate for the rapid growth of water demand, as water use is on a continuously upward trajectory. Some researchers have predicted that water demand in China may climb from 610.7 km^3 in 2011 to 670 km^3 in 2020 (Zhang, 2013). Within that total demand, estimates suggest that industrial water demand will be roughly 165–170 km^3, a 13%–16% increase from 2011; agricultural water demand will likely be 380 km^3, a 2% increase from 2011; domestic water demand will be approximately 95 km^3, a 20% increase from 2011; and environmental water demand will be about 20–25 km^3, a 78%–123% increase from 2011. Although agriculture will still be the largest water user in China, the growth of agricultural water demand is expected to slow in the coming decades.

Moreover, the spatially and seasonally uneven distributions of water resources may be exacerbated as a consequence of climate change (Piao et al., 2010). Northern China has more than 60% of the country's land, and 50% of the population, but only 18% of the water resources. Groundwater over-exploitation in northern China will cause the decline of the water table, leading to an increase in the energy needed to extract groundwater. Besides, groundwater contamination is trending to expand from point sources to non-point sources. China does not have an extensive groundwater remediation program, but is planning to establish a comprehensive groundwater monitoring network by 2020. The uneven distribution of water resources, increasing demand in water-scarce regions and serious groundwater contamination will likely accelerate the development of nontraditional water supplies (recycled water, desalinated water and large inter-basin water transfer projects). Given these trends, it is likely that energy demand in the water sector will continuously increase.

Water conservation strategies

The Chinese central government published the "Opinions of the State Council on the Implementation of the Most Stringent Water Resources Management" in 2012, which set mandatory targets (also known as the Three Red Lines) for national water use, water use efficiency and water pollution control by 2030 (http://www.gov.cn/zwgk/2012-02/16/content_2067664.htm). According to the targets, the upper limit of total water use is 700 km^3 by 2030, and the effective utilization coefficient of irrigation water, which is the ratio of water used by crops to the total irrigated water during an irrigation event, must be improved from the current 0.5 to 0.6. Agricultural irrigation has the greatest potential to reduce energy use through water saving because it accounts for 60% of total water use in China. If an irrigation utilization coefficient of 0.6 is achieved through better cropping structure and irrigation modes, then roughly 63 km^3 of irrigation water can be saved. This would reduce the energy required for water in agriculture by 17%. Another important issue is the leakage in the public water supply pipe network. In 2011, the water leakage rate was estimated to be 17%. This means that 17% of the energy used for water extraction, treatment and distribution was lost. Reducing the leakage rate could lead to energy savings.

Shifting water supply structure

Shifting the structure of water supply is also a potential way to reduce water-related energy use. The analysis of this study indicates that energy intensity of water extraction and treatment varies significantly, depending on the water sources. An overview of China's main water transfer projects and related energy intensity for water transportation is shown in Table A1 (in the online supplemental data). Water diversion over long distances is highly energy-intensive, and the energy intensity varies with distance. The intensity of energy consumption associated with different water sources also varies in different regions. Thus, regional differences should be considered when choosing the appropriate energy-saving water supplies. Northern China uses more energy-intensive water supplies due to the lack of adequate surface water. Taking Qingdao City (in Shandong Province) as an example, energy intensity of water provision (including water extraction and treatment) in this region is in the order of seawater desalination (5.9 kWh/m^3), inter-basin water transfer from the Yellow River (1.31 kWh/m^3), recycled water (0.82 kWh/m^3), groundwater supply (0.62 kWh/m^3) and local surface water supply (0.43 kWh/m^3), as discussed in the earlier section and Table A1. However, with the declining groundwater table, the energy needed for groundwater supply may become even higher than that for recycled water. Increasing recycled water use has the potential to save energy, particularly given that the current rate of wastewater use in the country is much lower than that of developed countries.

In addition to saving energy, water recycling may have two other benefits. First, a case study on urban water resources in Qingdao has indicated that developing water recycling could improve water security and reduce carbon emissions (Fu & Zhong, 2014). Second, in some instances energy can be recovered from wastewater treatment processes, and this recovered energy can be used to offset the electricity consumption. Wastewater contains energy in the form of potential energy, thermal energy and chemically bound energy. The potential energy depends on the topographically available head or difference in height. Thermal energy is a function of the temperature of the wastewater. The chemically bound energy results from its carbon content, which can be utilized through anaerobic digestion with biogas utilization and biosolids incineration with electricity generation (United Nations World Water Assessment Programme, 2014). A case study estimated that anaerobic digestion could save 628–4940 million kWh annually in the US (Stillwell, Hoppock, & Webber, 2010). There is still much room to develop the strategy for energy recovery from wastewater in China. It should be noted, however, that presently one constraint facing recycled water in general is the lack of pipe networks for wastewater collection in many urban areas.

Improving technology and infrastructure in both water and energy sectors

The energy flow analysis shows that the energy use for water pumping represents a majority of the total energy use in the water sector. Here water pumping includes not only groundwater pumping but also lifting and delivering surface water using pump stations. Improving pumping efficiency is essential for reducing energy use. In addition, energy intensity for water treatment (desalination and wastewater treatment) can vary significantly for different technologies. Therefore, improving technology, although costly in the short run, will be an effective way to reduce energy consumption.

Regarding electricity provision, there is also potential to reduce the transportation and distribution loss in electricity networks used for water pumping and treatment. In the past, China has emphasized the construction of power plants and neglected the construction of power supply networks. In developed countries, the average ratio of investment in power stations versus power networks is 5:5 or even 6:4, while the ratio in China is 4:6 (Gao & Luo, 2009). Under-investment in power utility networks has resulted in power loss during the distribution stage. In some rural regions, transmission and distribution losses can reach up to 30%. In recent years, China has been developing rural electric networks and increasing the investment in power network construction, which will greatly help with energy saving.

Challenges for policy reform

Both water and energy policy makers should give water conservation higher priority. Although China has gradually paid more attention to water and energy conservation, there are still many challenges ahead in water pricing, coordination amongst the agencies responsible for water and energy resources management, and infrastructure construction. Reasonable water pricing is an effective way to promote water conservation and indirectly to reduce energy use. China has applied various economic instruments for setting water prices, such as uniform volumetric tariff, two-part tariff and increasing block tariff. But water prices in China still do not reflect the real cost of producing and supplying water (SBC Energy Institute, 2014; Zhong & Mol, 2010). In uniform volumetric tariff, all units of water used are charged at the same rate, independent of the total consumption. Two-part tariffs include a monthly fixed-charge component plus a variable charge depending on the volume of water consumed. Under the increasing block tariff scheme, users pay a higher price for increasing consumption levels until the highest block of consumption. Only 133 of the 484 cities in China (less than 30%) had implemented increasing block tariffs by the end of 2013. Most cities are using uniform volumetric tariff. Though it is a water-scarce country, China charges a much lower water tariff compared to water-rich countries such as the UK (Zhong & Mol, 2010). Complicating any reform efforts, water governance in China is divided amongst several ministries and commissions, creating a very fragmented structure. This makes the practice of water conservation inefficient and ineffective (SBC Energy Institute, 2014). Meanwhile, water resource planning rarely incorporates energy considerations. Improving coordination amongst resources management agencies should be emphasized to promote both water and energy conservation. Beyond the regulatory environment, water supply structures and water engineering facilities need to be upgraded and greatly improved. For example, one constraint on recycled water utilization in China is the lagging utility pipe network construction for wastewater collection. Additionally, the high capital cost of adopting desalinated water and high-quality recycled water also represents an obstacle for shifting water supply structure.

Conclusions

This study collected and compiled data from multiple sources to calculate the energy use in China for water extraction, treatment, and distribution, as well as wastewater treatment and reuse. Spatial distributions of energy use by different water supplies were mapped at the provincial level to quantify regional differences. The results show that, for the main processes considered, water utilization consumes 193.5 TWh of electricity, or about 4% of the total national electricity consumption. The highest percentage of energy consumption for water is attributed to the water provision process (64%), especially for surface water provision (79.4TWh), due to the large amount of surface water use and inter-basin water transfer projects. From the perspective of water users, agricultural water provision consumes the highest proportion of energy. If wastewater treatment is taken into consideration, the industrial sector becomes the largest energy consumer.

It is essential to consider not only the availability and quality of water sources but also the energy use and carbon emissions of different types of water provision in order to develop a sustainable, low-carbon water supply, and to optimize the allocation of water resources. Increasing water demand is a driving factor for more energy-intensive water supplies in China. Thus, water conservation strategies are effective ways to save energy. From an energy point of view, recycled water use has potential to save energy because of its relatively low energy intensity compared to alternative water sources like desalination and long-distance water transfers. Moreover, new technological solutions (such as combined renewable energy and desalination plants, combined renewable energy and pump stations, and energy recovery from wastewater) offer promising ways to save energy.

The findings of this study could be improved with more site-specific data. First, the current analysis ended at the stage of water provision (including wastewater treatment) and did not consider energy for end use such as water heating. Second, regional and national averages were used to calculate the energy required for groundwater pumping and water treatment. More granular census data for both these processes would allow more accurate estimation of energy used by the water sector. Third, the energy embedded in the construction of water infrastructure was not accounted for in this study. Further studies that consider the energy used in the construction of water infrastructure would refine the results.

Last, this study focused on energy use for water; however, it is equally important to explore the water use for energy production. With the rapid increase of population and economic development in China, ensuring water and energy security simultaneously is becoming an unprecedented challenge. Moving forward, comprehensive studies that explore all aspects of the water–energy nexus for integrated and spatially explicit planning would help improve the sustainability and prosperity of the country.

Acknowledgements

The authors are grateful to Christopher Napoli and the other two anonymous reviewers for their detailed comments, which have significantly improved the presentation of this work.

Disclosure statement

No potential conflict of interest was reported by the authors.

Funding

Support for this work was provided by the National Natural Science Foundation of China [Grants 91225301, 41271032 and 91025019].

References

Al-Karaghouli, A. & Kazmerski, L. L. (2013). Energy consumption and water production cost of conventional and renewable-energy-powered desalination processes. *Renewable and Sustainable Energy Reviews, 24*, 343–356. doi:10.1016/j.rser.2012.12.064.

Bartos, M. D. & Chester, M. V. (2014). The conservation nexus: Valuing interdependent water and energy savings in Arizona. *Environmental Science & Technology, 48*, 2139–2149. doi:10.1021/es4033343.

Buonomenna, M. G. (2013). Membrane processes for a sustainable industrial growth. *RSC Advances, 3*, 5694–5740. doi: 10.1039/C2RA22580H.

China Geological Environmental Monitoring Institute (2012). *China groundwater level yearbook of geological environmental monitoring*. Beijing: China Land Press.

China National Census for Water (2013). *Bulletin of first national census for water*. Beijing: China Waterpower Press.

China Urban Water Association (2012). *China urban water supply yearbook*. Beijing: China Urban Water Association Press.

CWRB. (2012). *China water resources bulletin*. Beijing: Ministry of Water Resources of the People's Republic of China. Retrieved from the website: http://www.mwr.gov.cn/zwzc/hygb/szygb/qgszygb/201212/P020121217531238597482.doc.

Fu, X. & Zhong, L. (2014). *Water-Energy nexus of urban water systems for Chengdu's low-carbon blueprint*. Beijing: World resources Institute.

Gao J. (2012). *The nexus of water resource and electricity production in China*. (Master's thesis), Tianjin University, Tianjin.

Gao, P. & Luo, G. (2009). Problems in development of electrical power in rural China. *Asian social science, 5*, P27. doi: 10.5539/ass.v5n12p27

Gude, V. G. (2011). Energy consumption and recovery in reverse osmosis. *Desalination and Water Treatment, 36*, 239–260. doi:10.5004/dwt.2011.2534

Gude, V. G., Nirmalakhandan, N., & Deng, S. (2010). Renewable and sustainable approaches for desalination. *Renewable and Sustainable Energy Reviews, 14*, 2641–2654. doi:10.1016/j.rser.2010.06.008

Guo, Y., Wang, X., & Zhou, Z. (2012). National survey of urban sewage reuse in China. *Enviromental Science, 33*, 3881–3884.

Hu, G., Ou, X., Zhang, Q., & Karplus, V. J. (2013). Analysis on energy–water nexus by Sankey diagram: The case of Beijing. *Desalination and Water Treatment, 51*, 4183–4193. doi:10.1080/19443994.2013.768038.

Kahrl, F. & Roland-Holst, D. (2008). China's water–energy nexus. *Water Policy, 10*, 51. doi:10.2166/wp.2008.052.

Karimi, P., Qureshi, A. S., Bahramloo, R., & Molden, D. (2012). Reducing carbon emissions through improved irrigation and groundwater management: A case study from Iran. *Agricultural Water Management, 108*, 52–60. doi:10.1016/j.agwat.2011.09.001.

Li, X. Y., Gong, J. D., & Wei, X. H. (2000). In-situ rainwater harvesting and gravel mulch combination for corn production in the dry semi-arid region of China. *Journal of Arid Environments, 46*, 371–382. doi:10.1006/jare.2000.0705.

Li, C., Wang, Y., & Qiu, G.- Y. (2013). Water and energy consumption by agriculture in the minqin oasis region. *Journal of Integrative Agriculture, 12*, 1330–1340. doi:10.1016/s2095-3119(13)60542-0.

Liu, J., & Yang W. (2012). Water management. water sustainability for China and beyond. *Science, 337*, 649–650. doi: 10.1126/science.1219471

Liu, J., Zang, C., Tian, S., Liu, J., Yang, H., Jia, S., & Zhang, M. (2013). Water conservancy projects in China: Achievements, challenges and way forward. *Global Environmental Change, 23*, 633–643. doi:10.1016/j.gloenvcha.2013.02.002.

Ma, L., Allwood, J. M., Cullen, J. M., & Li, Z. (2012). The use of energy in China: Tracing the flow of energy from primary source to demand drivers. *Energy, 40*, 174–188. doi:10.1016/j.energy.2012.02.013.

Margat J., & Van der Gun J. (2013). *Groundwater around the world: A geographic synopsis*. Boca Raton: CRC Press.

MOHURD. (2012). *China urban water supply bulletin*. Beijing:Ministry of Housing and Urban-Rural Development of the People's Republic of China. Retrieved from the website: http://www.mohurd.gov.cn/xytj/tjzljsxytjgb/w02012052419769781411330032.doc.

NBS (2012a). *China energy statistical yearbook*. Beijing: National Bureau of Statistics of the People's Republic of China.

NBS (2012b). *China statistical yearbook on environment in 2011*. Beijing: National Bureau of Statistics of the People's Republic of China.

NBS. (2012c). *China statistical yearbook*. Beijing: National Bureau of Statistics of the People's Republic of China.

NDRC. (2012). *The 12th five-year plan of national rural drinking water safety project*. Beijing: The National Development and Reform Commission. Retrieved from the website: http://www.jsgg.com.cn/Files/PictureDocument/20131029133751645554645892.pdf.

Piao, S., Ciais, P., Huang, Y., Shen, Z., Peng, S., Li, J., & Fang, J. (2010). The impacts of climate change on water resources and agriculture in China. *Nature, 467*, 43–51. doi:10.1038/nature09364.

Plappally, A. K., Lienhard, V. J. H. (2012). Energy requirements for water production, treatment, end use, reclamation, and disposal. *Renewable and Sustainable Energy Reviews, 16*, 4818–4848. doi:10.1016/j.rser.2012.05.022.

Power, C. (2012). *China electric power yearbook*. Beijing: China Electric Power Press.

Qin, Y., Curmi, E., Kopec, G. M., Allwood, J. M., & Richards, K. S. (2015). China's energy-water nexus – assessment of the energy sector's compliance with the "3 Red Lines" industrial water policy. *Energy Policy, 82*, 131–143. doi:10.1016/j.enpol.2015.03.013.

Qiu J. (2010). China face up to groundwater crisis. *Nature, 466*, 308. doi:10.1038/466308a.

Reffold E., Leighton F., Choudhury F., & Rayner P. S. (2008). *Greenhouse gas emissions of water supply and demand management options*. Bristol: Enviromental Agency. Retrieved from https://www.gov.uk/government/uploads/system/uploads/attachment_data/file/291728/scho0708bofv-e-e.pdf.

Rothausen, S. G. S. A. & Conway, D. (2011). Greenhouse-gas emissions from energy use in the water sector. *Nature Climate Change, 1*, 210–219. doi:10.1038/nclimate1147.

Sankey H. (1898). Introductory note on the thermal efficiency of steam-engines. *Minutes of Proceedings of The Institution of Civil Engineers*, 278–283.

SBC Energy Institute. (2014). Introduction to the water and energy challenge. Retrieved from: https://www.sbc.slb.com/~/media/Files/SBC%20Energy%20Institute/SBC%20Energy%20Institute_Introduction%20to%20the%20Water%20and%20Energy%20Challenge.ashx.

Schmidt, M. (2008). The sankey diagram in energy and material flow management. *Journal of Industrial Ecology, 12*, 82–94. doi:10.1111/j.1530-9290.2008.00004.x

Shah, T. (2009). Climate change and groundwater: India's opportunities for mitigation and adaptation. *Environmental Research Letters, 4*, 035005. doi:10.1088/1748-9326/4/3/035005.

Shao, X., Wang, H., & Wang, Z. (2003). Interbasin transfer projects and their implications: A China case study. *International Journal of River Basin Management, 1*, 5–14. doi:10.1080/15715124.2003.9635187.

SOA. (2013). *National sea water use report in 2012*. Beijing: State Oceanic Administration People's Republic of China.

Stillwell, A. S., Hoppock, D. C., & Webber, M. E. (2010). Energy recovery from wastewater treatment plants in the United States: A case study of the energy-water nexus. *Sustainability, 2*, 945–962. doi:10.3390/su2040945.

United Nations World Water Assessment Programme. (2014). *The United Nations world water development report 2014: Water and energy*. Paris: UNESCO. http://unesdoc.unesco.org/images/0022/002257/225741E.pdf.

U.S. Energy Information Administration. (2014). Retrieved from the website: http://www.eia.gov/countries/analysisbriefs/China/china.pdf.

Wang, Y., Xie, Z., Malhi, S. S., Vera, C. L., & Zhang, Y. (2009). Effects of rainfall harvesting and mulching technologies on water use efficiency and crop yield in the semi-arid Loess Plateau, China. *Agricultural Water Management, 96*, 374–382. doi:10.1016/j.agwat.2008.09.012.

Wang, J., Rothausen, S. G. S. A., Conway, D., Zhang, L., Xiong, W., Holman, I. P., & Li, Y. (2012). China's water–energy nexus: Greenhouse-gas emissions from groundwater use for agriculture. *Environmental Research Letters, 7*, 014035. doi:10.1088/1748-9326/7/1/014035.

Wei, C. & Shen, M. (2009). Current progresses and new trend of research on energy efficiency: A literature review. *Journal of Zhejiang University (Humanities and Social Sciences), 39*, 55–63.

Wen, H., Zhong, L., Fu, X., & Spooner, S. (2014). *Water energy nexus in the urban water source selection_a case study from Qingdao*. Beijing: Water Resources Institute.

Zhang L. (2013). Overlook of China's water demand in the next ten years. *Development Research*, 12–18

Zhong, L. & Mol, A. P. (2010). Water price reforms in China: Policy-making and implementation. *Water Resources Management, 24*, 377–396.

Zou, X., Li, Y. E., Li, K., Cremades, R., Gao, Q., Wan, Y., & Qin, X. (2013). Greenhouse gas emissions from agricultural irrigation in China. *Mitigation and Adaptation Strategies for Global Change, 20*, 295–315. doi:10.1007/s11027-013-9492-9.

Index

For Product Safety Concerns and Information please contact our EU representative GPSR@taylorandfrancis.com
Taylor & Francis Verlag GmbH, Kaufingerstraße 24, 80331 München, Germany

www.ingramcontent.com/pod-product-compliance
Ingram Content Group UK Ltd.
Pitfield, Milton Keynes, MK11 3LW, UK
UKHW050753180926
475701UK00007B/1032
* 9 7 8 0 3 6 7 2 3 4 0 8 9 *